Martin Hochberg

A high-voltage pulsed power modulator for fast-rising arbitrary waveforms

Karlsruher Forschungsberichte aus dem
Institut für Hochleistungsimpuls- und Mikrowellentechnik

Herausgeber: Prof. Dr.-Ing. John Jelonnek

Band 14

A high-voltage pulsed power modulator for fast-rising arbitrary waveforms

by

Martin Hochberg

KIT Scientific Publishing

Karlsruher Institut für Technologie
Institut für Hochleistungsimpuls- und Mikrowellentechnik

A high-voltage pulsed power modulator
for fast-rising arbitrary waveforms

Zur Erlangung des akademischen Grades eines Doktor-Ingenieurs
von der KIT-Fakultät für Elektrotechnik und Informationstechnik des
Karlsruher Instituts für Technologie (KIT) genehmigte Dissertation

von M.Sc. Martin Christian Hochberg

Tag der mündlichen Prüfung: 17. Juli 2019
Hauptreferent: Prof. Dr.-Ing. Georg Müller
Korreferent: Prof. Dr.-Ing. Thomas Leibfried

Impressum

KIT Scientific Publishing

Karlsruher Institut für Technologie (KIT)
KIT Scientific Publishing
Straße am Forum 2
D-76131 Karlsruhe

www.ksp.kit.edu

Print on Demand 2020 – Gedruckt auf FSC-zertifiziertem Papier

ISSN 2192-2764
ISBN 978-3-7315-0958-5
DOI 10.5445/KSP/1000097344

Foreword of the editor

One of the main research interests at the Institute of Pulsed Power and Microwave Technology (IHM), Karlsruhe Institute of Technology (KIT), is the enhancement of metal surface properties by means of pulsed electron beam treatment. The required homogeneous high-current electron beam is created in a vacuum triode configuration (GESA). Its acceleration voltage is supplied by a driving pulsed power source capable of delivering a high-voltage pulse of 120 kV at 600 A for a duration of up to 30 µs with a voltage rise rate higher than 10^{12} V/s. Prior to this work, no semiconductor-based pulsed power source could meet these requirements – therefore a spark-gap based generator was used with inherent drawbacks for the process as it does not allow for a direct control of the pulse parameters.

For maturing the process to industry level, Dr.-Ing. Martin Hochberg developed, constructed and tested a semiconductor-based generator meeting the above-mentioned challenging pulse parameters for the first time.

The chosen modular concept of the generator with a stage voltage of 1 kV allows for an adjustability of the output voltage in a step width below 1 ‰. To overcome the limitations with respect to switching speed for cost-efficient IGBTs, he investigated a "gate-boosting" circuit allowing a tremendous increase in switching speed for voltage-controlled power semiconductors. The resulting fast switching elements in combination with a low-inductance capacitor arrangement allowed for current rise rates exceeding 10 kA/µs at switching times below 50 ns for a single stage. By controlling each stage locally with a microprocessor, he could simplify the main control circuitry and shield the sensitive electronics successfully against electro-magnetic interference. A new optical bus system connecting the stages allows for a simple assembly process and a fast turn-off in case of system errors. In his design, Dr.-Ing. Martin Hochberg included a fast current measurement on each stage to interrupt pulse execution in case of over-current conditions and validated it in different fault scenarios. Using 3D electro-magnetic simulations, the arrangement's inductance and stray capacitance were considered during the design process to ensure the rise time requirement could be met prior to the construction of the generator. In the final assembly, the unprecedented pulse parameters of the new generator were validated – with a rise time exceeding the requirement by 20 %. Additionally, Dr.-Ing. Martin Hochberg demonstrated the topology's unique capability of achieving a flat-top pulse by subsequent activation of spare stages. The results of this work therefore present a key step towards enhanced control of the electron beam parameters, enabling its use for a broad range of applications.

Zusammenfassung

Die Verwendung gepulster Elektronenstrahlen zur thermischen Oberflächenbehandlung bietet Vorteile in vielen Bereichen der Energieproduktion, da durch den Prozess eine Verbesserung der Verschleiß- und Korrosionsbeständigkeit metallischer Werkstücke erzielt werden kann. Für den erfolgreichen Einsatz einer gepulsten Elektronenstrahl-Anlage (GESA) ist dabei der verwendete Hochleistungsimpuls-Generator entscheidend, der den benötigten Hochleistungsimpuls von 120 kV, 600 A bei einer Pulslänge von bis zu 100 µs bereitstellt. Der bisher verwendete funkenstreckenbasierte Generator limitiert durch seine begrenzte Flexibilität (rechteckige Ausgangsspannungsform, schwer einstellbare Pulslänge und Amplitude) die Anwendung stark. Um eine Weiterentwicklung der Technologie zur vollständigen Industriereife zu ermöglichen, wurde im Rahmen dieser Arbeit erstmalig ein halbleiterbasierter Hochleistungsimpuls-Generator entwickelt, gebaut und getestet, der zur Ansteuerung einer GESA-Anlage verwendet werden kann. Er zeichnet sich durch die feine Modulierbarkeit seiner Ausgangsspannung in Stufen von unter 1 % der maximalen Ausgangsspannung während des Pulses aus, welche durch den modularen Aufbau mit einer Stufenspannung von bis zu 1000 V erreicht wurde. Die Kombination aus den Pulsparametern und der Notwendigkeit eines schnellen Spannungsanstiegs von 10^{12} V/s am Ausgang des Generators stellte eine besondere Herausforderung dar, die bisher nicht mit konventionell angesteuerten Leistungshalbleitern realisiert wurde. Zur drastischen Erhöhung der Schaltgeschwindigkeit spannungsgesteuerter Leistungshalbleiter wurde hierzu eine neuartige "Gate-boosting" Schaltung entwickelt. Weiterhin wurde durch einen speziellen Platinenentwurf eine niederinduktive Verbindung sowohl der Pulskondensatoren zu den Schaltelementen als auch der Platinen untereinander erreicht. Um einen schnellen Spannungsanstieg zu gewährleisten, wurden neben der Induktivität des Aufbaus auch die zu erwartenden Streukapazitäten mittels 3D elektro- und magnetostatischen Simulationen analysiert und im Entwurf berücksichtigt. Eine erhebliche Reduktion der Komplexität der Kontrolleinheit konnte durch die dezentrale Ansteuerung der Schaltelemente durch Mikroprozessoren und schnelle Logik lokal auf der Platine erzielt werden. Zugleich wird dadurch eine schnelle Reaktion des Generators auf Überstromfehler ermöglicht. Die effiziente Kommunikation aller Stufen untereinander gelingt durch die Implementierung eines spezialisierten optischen Bussystems, welches die Synchronisierung aller Stufen mit einer Genauigkeit von unter 11 ns erlaubt. Es konnte gezeigt werden, dass das zum Schutz der Elektronik entworfene Schirmkonzept eine Beeinflussung durch die beim Betrieb des Generators auftretenden elektromagnetischen Störungen verhindert. Neben neuen technischen Ansätzen enthält das entwickelte Design auch mehrere Konzepte zur starken Reduktion der Projektkosten. Nach erstmaliger Validierung der Generatorparameter im kleinen Maßstab von 8 Platinen wurde der vollständige Generator mit 149 Stufen erfolgreich aufgebaut und mit zuvor unerreichten Pulsparametern in Betrieb genommen. An einer der GESA-Anlage ähnlichen Last konnte die geforderte Spannungssteilheit bei voller Ausgangsspannung von 120 kV um 20 % übertroffen werden. Es wurde gezeigt, dass durch sukzessives Zuschalten von Ersatzstufen die Spannungsabnahme der Pulskondensatoren effektiv kompensiert werden kann. Der neu entwickelte Generator steht nun zur Integration in das GESA-Versuchsfeld zur Verfügung.

Abstract

Pulsed electron beam treatment for thermal surface modification offers new possibilities in many fields of energy technology, as it allows the enhancement of the wear and corrosion resistance of metals. A critical element of the process is the employed pulsed power modulator driving the pulsed electron beam device (GESA). It supplies the necessary high-power pulse with a total output voltage of 120 kV and a total current of 600 A for a duration of up to 100 µs. The currently employed spark-gap based modulator restricts the applications of the process due to its limited flexibility in terms of output voltage shape, amplitude and duration. To enable a maturing of the technology to large-scale industry level, a semiconductor-based pulsed power modulator capable of driving a GESA device was developed, build and tested for the first time in the scope of this work. It allows for a step-wise arbitrary output waveform of up to 120 kV and a maximum current of 600 A. By selecting a modular concept and a stage voltage of 1000 V, the output voltage can be modulated in fine steps below 1 % of the maximum output voltage. Besides the unique pulse parameters, the required fast rise rate of the output voltage of 10^{12} V/s posed a special challenge, as it has not been realized with presently available power semiconductors using standard gate-drive technologies. To overcome this limitation, a new "gate-boosting" circuit was developed, allowing a tremendous increase in switching speed for commercial voltage-controlled power semiconductors. Additionally, the circuit inductance was kept low using a special board layout for reduced inductance between the pulse capacitors and the pulse switches on the one hand, as well as between adjacent stages on the other hand. Together with the parasitic inductive properties of the assembly, the parasitic capacitances were analyzed based on 3D electro- and magneto-static simulations and considered in the design for meeting the voltage rise time requirement. The employed decentralized control of the pulse switches by microprocessors and fast logic locally on each board drastically reduces the complexity of the main control unit while ensuring a fast reaction time of the modulator to overcurrent conditions. An efficient communication between all stages is achieved by implementing a specialized optical bus system, enabling a synchronization of all stages within 11 ns. The designed electro-magnetic shielding for the sensitive electronics on each stage was tested successfully. Complementing the new solutions for technical challenges, the presented work implements approaches for drastically cutting the project costs. After a successful validation of the design parameters in a proof-of-concept assembly consisting of eight stages, the full-scale modulator was assembled and tested with unprecedented pulse parameters. Connected to a load similar to the GESA device, the required output voltage rise rate at full output voltage of 120 kV could be exceeded by 20 %. The subsequent activation of spare stages proved efficient in counteracting the pulse capacitor voltage droop. The modulator is now ready for integration into the GESA experiment.

Table of Contents

List of Abbreviations

ADC	Analog-to-digital converter
Be	Beryllium
Bi	Bismuth
CAD	Computer-aided design
CPLD	Complex programmable logic device
CPU	Central processing unit
Cu	Copper
DC	Direct current
DUT	Device-under-test
EMI	Electro-magnetic interference
FPGA	Field programmable gate array
FS	Field-stop
GaN	Gallium-nitride
GESA	Gepulste Elektronenstrahl-Anlage, German for "Pulsed electron beam device"
HVPS	High-voltage power supply
IGBT	Insulated-gate bipolar transistor
IHM	Institut für Hochleistungsimpuls- und Mikrowellentechnik

KIT	Karlsruher Institut für Technologie
LC	Inductor-capacitor combination
LED	Light-emitting diode
LPF	Low-pass filter
LTD	Linear transformer driver
MMC	Modular multilevel converter
MOSFET	Metal-oxide-semiconductor field-effect transistor
NPT	Non-punch-through
Pb	Lead
PCB	Printed circuit board
PSM	Pulse Step modulator
RC	Resistor-capacitor combination
RF	Radio-frequency
RL	Resistor-inductor combination
RLC	Resonant circuit with resistive, inductive and capacitive properties
RMS	Root-mean-square
SEM	Scanning electron microscope
Si	Silicon
SiC	Silicon carbide
SMD	Surface-mounted device
SOA	Safe operating area

TC	Timer-counter
TTL	Transistor-transistor logic
TVS	Transient voltage suppressor diode
ZCS	Zero-current switching
ZVS	Zero-voltage switching

1 Introduction

1.1 Motivation

The surface modification of metal targets by means of intense pulsed electron beams (GESA process) is one of the main research interests at the Institute for Pulsed Power and Microwave Technology (IHM), Karlsruhe Institute of Technology (KIT). The high energy deposition in the topmost surface layer of the sample by the intense electron beam causes the surface to melt – the associated high heating and cooling rates lead to a restructuring of the surface layer. Thereby, the process can be used to either enhance the surface properties in terms of surface hardness and wear resistance or even change the surface composition by alloying thin films of material into the surface, both without affecting the bulk material beneath it [1]. Especially for structural materials in contact with liquid metals (as used in concentrated solar power plants or accelerator-driven transmutation reactors [2], [3]) an increased corrosion resistance can be achieved [4]. The electron beam is created using the GESA (German acronym "Gepulste ElektronenStrahl-Anlage", "pulsed electron beam device"), a high-current electron accelerator for large area surface treatment, originally developed in cooperation with the Efremov Institute St. Petersburg, Russia, starting in 1996. The device operates in pulsed mode with a pulse length of up to several tens of microseconds. It requires an acceleration voltage of up to 120 kV at a beam current of up to 600 A. The high peak power in the range of 72 MW is supplied by a specialized pulsed power supply. Currently, the device is equipped with a spark-gap based power supply with very limited flexibility. Due to its design, the output voltage is rectangular and any change of pulse length or pulse amplitude requires a physical manipulation of components. Recent progress in the development of power semiconductor devices offers new possibilities for their use in high-voltage pulsed power sources with respect to active pulse shaping. For a deeper understanding of the processes governing the GESA device physics and, subsequently, enabling

a maturing of the technology to industry level, it is the scope of this thesis to develop, build and test a new type of semiconductor-based pulsed power modulator capable of active pulse shaping. The requirements for future experiments include a flexible output voltage of up to 120 kV, that can be modulated with a step-size below 1 % of the maximum voltage. At a nominal current of 250 A, the maximum pulse length should be 100 µs. The most important figure of merit is the output voltage rise rate when connected to the GESA device, specified to 10^{12} V/s.
While specialized pulsed power sources meeting one of the requirements or another have been presented before, the combination of the requested parameters renders this project very challenging. On the one hand, the GESA device requires a very fast output voltage rise rate, on the other hand it demands for a comparably long high-current pulse. Both requirements are conflicting, since long pulses require big energy storage capacitors which, in turn, increase the circuit inductance (for a given volume) and therefore slow down the output voltage rise rate. Currently, cost-efficient power semiconductor devices meeting the speed requirement for this project are not available. In the presented design, the fine step-size in combination with the fast rise rate requires many individual switching elements, all controlled independently from each other at minimal financial and fabrication effort. Due to the high output voltage and current, the sensitive electronics need to be shielded appropriately against electro-magnetic interference. During the design process, all parasitic circuit parameters such as stray inductance and stray capacitance need to be simulated to meet all the requirements. Finally, the project has to be planned according to the manufacturing capabilities of the institute's workshop and within a limited budget.
The presented thesis describes the design process to meet all the above-mentioned requirements and their validation with first measurements on the full-scale version of the modulator.

1.2 Structure of this thesis

In this work, the development of a new semiconductor-based pulsed power modulator for the GESA device is presented. In contrast to the former generator, the voltage amplitude, pulse length, and pulse shape of the output pulse are controlled by software.

After an overview over the GESA experiment and the deduction of the pulsed power source requirements, chapter 2 describes the generator design study. The selected generator topology is presented and compared to other approaches found in literature. As no generator for the required set of pulse parameters has been demonstrated before, the resulting challenges with respect to the circuit components are outlined.
A very strict requirement of the device is an output voltage rise rate in the order of 10^{12} V/s. As this parameter cannot be met in a cost-efficient way with currently available power semiconductors, chapter 3 presents a new "gate-boosting" circuitry, which allows the speed-up of commercial devices by a factor of up to eight.
The investigated topology is modular and consists of identical stages connected in series to achieve the required full output voltage. Chapter 4 describes the design and validation process of one stage and covers all important circuit elements: selection and parallel operation of IGBTs using a gate-boosting circuit, low-inductive energy storage, switching signal generation, and related protection circuitry.
In chapter 5, the combined operation of up to eight stages (maximum output voltage of 8 kV) in a generator arrangement is presented. As interface between the user (PC) and the generator, a control unit has been developed. For reducing its complexity, the stages communicate via a new optical bus system, allowing for a low-jitter synchronization of all stages. All features of the generator, namely the arbitrary waveform generation at full output voltage, maximum load current (600 A), and fast rise time as well as an efficient overcurrent protection scheme have been validated in the full control chain.
Finally, the full-scale generator has been constructed and operated in a test-bed. The measured output pulse shape is in good agreement with the simulations and shows a voltage rise time even exceeding the required 10^{12} V/s by around 20 %. In addition, active pulse shaping could be demonstrated to achieve droop compensation at full output voltage.

2 Generator design study

For high-voltage, high-current pulsed power sources, several architectures are known. The generator topology investigated in this work is the semiconductor-based Marx generator. It was chosen due to its scalable design, allowing for achieving all pulse parameter requirements at reasonable costs. The following chapter starts with a short overview over the GESA device and deduces the required pulse parameters. Consecutively, it presents the selected generator architecture and compares it to previous work described in literature. Subsequently, the design study covering the pulsed power components, the logic and auxiliary circuitry necessary to achieve all requirements in the presented architecture is discussed. Selected results of this work presented in the following chapter have already been published as [5].

2.1 Design parameters

In the Institute for Pulsed Power and Microwave Technology (IHM) at the Karlsruhe Institute of Technology, Germany, several devices exist for treatment of samples with a wide parameter range. Whereas the applications are similar for all devices, the pulse parameters differ significantly. A comprehensive overview is given in [6] and [7]. All values given in this work refer to the GESA I device, to which the generator-under-development will be connected.

2.1.1 GESA pulse parameters

The basic concept behind the GESA device is a vacuum tube in a triode configuration, consisting of a cathode, a control grid, and an anode. A schematic can be seen in Fig. 2.1a. A pulse generator delivers a negative high-voltage pulse of around −120 kV to the cathode. The cathode, consisting of an array of carbon fiber bundles, ignites a plasma serving as electron emitter. By means of the potential difference between cathode and control grid, an intense electron beam of up to 600 A can be extracted from the plasma and is accelerated towards the anode. Several magnetic coils generate a guiding magnetic field in the device [8] to focus the beam through an opening in the anode and onto a metal target beneath it. In the photograph shown in Fig. 2.1b, the upper part of the image shows the cathode on high-voltage potential, whereas the lower part shows the treatment chamber and the required vacuum system on ground potential.

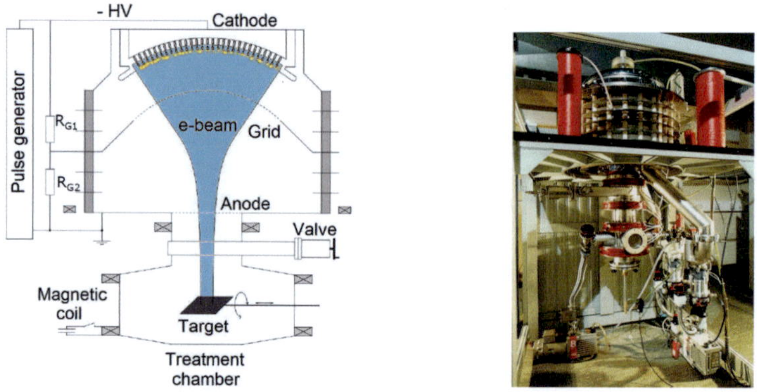

(a) **(b)**

Figure 2.1: Schematic (a) and photo (b) of the pulsed electron beam device (GESA).

The operating voltage for the cathode is in the range of 50...150 kV [9], whereas a value of 120 kV is preferred. The typical potential distribution and the resulting beam current are depicted in Fig. 2.2. A part of the beam current (in the range of 10 %) hits the grid and is deflected via the resistive network R_{G1} and R_{G2}, implementing a negative feedback – an increase in beam current will decrease the potential difference between cathode and grid. However, this mean of control has proven to be unsatisfactory, limiting the maximum achievable pulse length to 50 μs. For a better control of the process and future research on the plasma dynamics, the generator will be required to supply a maximum output voltage of 120 kV, adjustable in steps below 1 %.

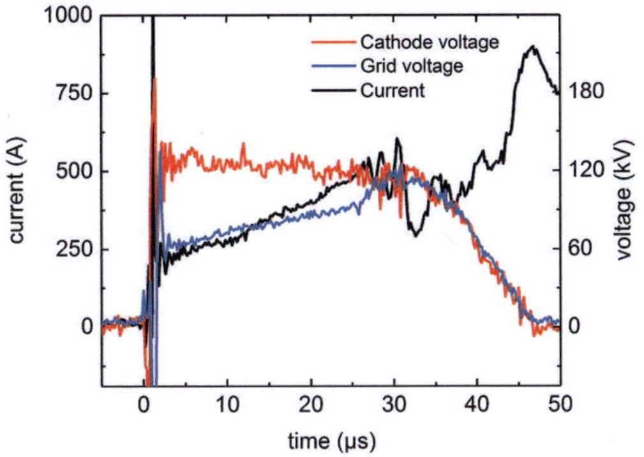

Figure 2.2: A typical pulse of the GESA device. The electron beam is extracted by the potential difference between cathode voltage and grid voltage.

Additionally, the topology has to allow for a direct control of the grid voltage. The nominal current of the generator is specified to 250 A with a maximum pulse duration of 100 µs. Due to the dynamic load behavior of the GESA device, a complete short may occur. Therefore an overcurrent protection has to be implemented if the beam current significantly exceeds 600 A. The maximum repetition rate will be one shot every 30 s and thereby comparable to the current design. This work assumes implicitly a deterministic and stable behavior of the process. As result, an open-loop control is implemented in the generator. However, the developed hardware includes the possibility of upgrading the generator to a closed-loop control system if this assumption proves wrong in the course of future experiments.

2.1.2 Homogeneous cathode ignition

As electron emitter, the GESA device uses a "multipoint explosive emission cathode". Depicted in Fig. 2.3a, the cathode is an arrangement of individual carbon fiber bundles. A magnification of one bundle is shown in the SEM image in Fig. 2.3b. Typically, field emission starts at electric fields of around 10^5 V/cm [10]. By using fibers with a diameter of 10 µm, the on-set of field emission can be reduced to a field strength of 10^2 V/cm [10] due to field enhancement. When the pulse generator (compare Fig. 2.1) turns on, the threshold for field emission is exceeded. The cold carbon fibers start to emit electrons near their Fermi level, whereas the replacement electrons from the pulse generator enter the carbon fibers at the highest unoccupied energy level. The energy difference between these states is dissipated causing a heating of the fiber ("Nottingham effect") [11]. With increasing temperature, electron emission is facilitated. The increasing electron current causes ohmic heating of the fiber, changing the emission mechanism to thermionic emission [12]. If the voltage rise is fast enough an evaporation of the emitting fiber tip [10] occurs within several nanoseconds ("explosive electron emission") [13]. The thereby formed plasma layer expands and acts as emission surface.

(a) The GESA cathode consists of an array of carbon fiber bundles, each with an individual resistor for current balancing.

(b) The tips of the individual carbon fibers have a diameter of 10 µm and serve as electron emission centers.

Figure 2.3: Photograph of the cathode arrangement with 700 individual carbon fiber bundles and scanning electron microscope (SEM) image of one bundle showing the individual fibers.

To foster a homogeneous cathode plasma ignition, a voltage rise in the order of 10^{12} V/s is required [10].
Whereas some applications are very sensitive against overvoltage [14], the GESA device was designed for voltages of 150 kV and is operated at nominal voltage of 120 kV. Any occurring overvoltage due to parasitic capacitances/inductances of the generator will increase the voltage rise rate and foster a homogeneous plasma ignition. Once the plasma has ignited, the drop in the cathode's impedance will dissipate the energy stored in the parasitic capacitances and, therefore, limit the duration of the overvoltage. Since the currently employed pulse generator cannot achieve the required rise time, a

peaking circuit is implemented [10]. To reduce the circuit complexity, the generator under development has to meet the demand for the voltage rise time across the cathode without the need of a peaking circuit. The figure of merit for the design of the generator used in this work is therefore the voltage rise rate from the beginning of the pulse (10 %, i.e. 10 kV) to 100 kV, regardless of a possible voltage overshoot.

2.1.3 Parasitic device properties

Apart from the inherent device physics, the mechanical dimensions of the device determine its behavior in an electric circuit. Especially for the voltage rise time calculations, the parasitic properties such as device inductance, and device capacitance are crucial.

(a) A photo of the GESA cathode and field grading rings in its shielding enclosure.

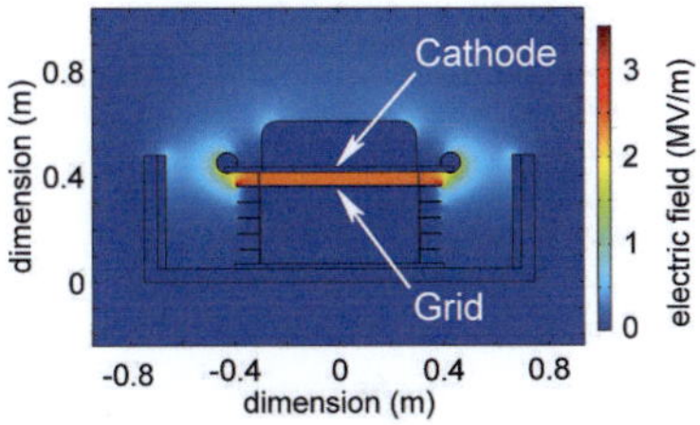

(b) Simplified 3D model and resulting electric field.

Figure 2.4: Photo and corresponding 3D model used for FEM static electric field simulations to estimate the cathode-to-ground stray capacitance.

Fig. 2.4a shows a photo of the GESA cathode in its lead-reinforced metal shielding enclosure. In order to avoid a time-consuming precise modeling of all internal components, a worst-case estimation was performed to derive the parasitic values of the device. Concerning the parasitic capacitance of the device, Fig. 2.4b illustrates static electric field simulations performed using COMSOL (AC-DC module). The outside of the device was modeled according to its physical dimensions, with the cathode simulated at a constant voltage of 120 kV. As worst-case scenario, the control grid was simulated to be on ground potential. From the electric field energy, the total resulting equivalent capacitance was calculated to 150 pF. A crude estimation concerning the device's parasitic inductance can be performed on basis of its geometric dimensions. When connected to the center of the cathode and the edge of the shielding enclose, the resulting current loop will have a total length of about 4 m. Due to the large diameter, the coupling between the conductors is not significant. Therefore, an approximate inductance of 10 nH/cm [15] can be assumed, yielding a total inductance of 4 µH.

2.2 The semiconductor-based Marx generator

The semiconductor-based Marx generator is a member of a group of topologies of a step-wise arbitrary voltage sources not limited to the field of pulsed power. Conceptually similar, but with different parameters and much slower switching speeds, are the Modular Multilevel Converter (MMC, [16], [17]) or the Pulse Step Modulator (PSM, [18]).

2.2.1 Topology

The Marx generator, a circuit patented 1923 by Erwin Marx [19], was originally invented to create high-voltage impulses for insulator testing.

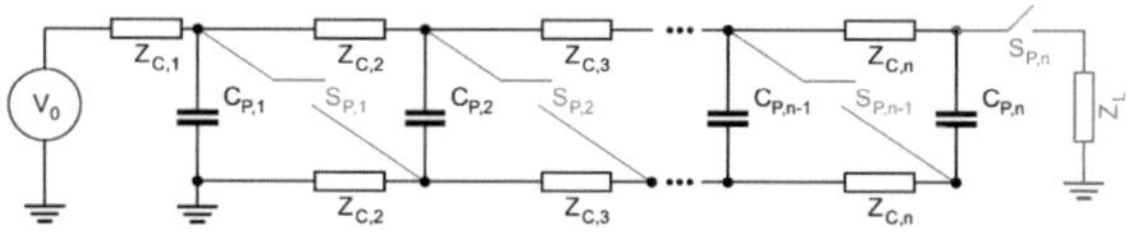

(a) During charging, all capacitors are connected in a parallel configuration.

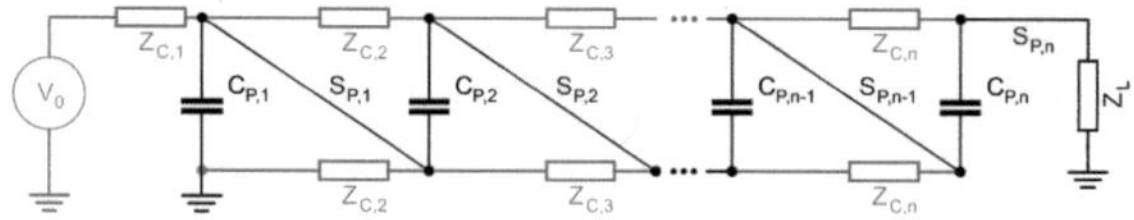

(b) By closing the pulse switches, the capacitors are arranged in a series configuration and discharge via the load impedance Z_L.

Figure 2.5: The Marx generator comprised of n identical cells, each consisting of a pulse capacitor C_P, a pulse switch S_P and a charging impedance Z_C for transient decoupling during the pulse.

The modular concept is illustrated in Fig. 2.5: each of the n stages consists of a pulse capacitor C_P, a charging impedance Z_C and a pulse switch S_P. Prior to pulse execution, all capacitors are connected in parallel via the charging impedances and charged to the same voltage V_0. When the pulse switches are closed, the capacitors are connected in a series arrangement causing a voltage multiplication to $n \cdot V_0$ across the load impedance Z_L. The charging impedances Z_C decouple adjacent stages during the discharge time of the capacitors, preventing a significant short circuit current. Since all stages are floating during the pulse, the maximum voltage each component has to withstand is solely the charging voltage V_0, even if the output voltage is n times higher. Thereby, the generator can by composed of very cost-efficient components. When erected, the presented circuit (Fig. 2.5b) effectively is an RLC circuit composed of the load impedance (circuit resistance and inductance) and the series connection of all pulse capacitors. Depending on the degree of damping, the pulse into the load can be over-damped, critically damped or oscillating. The same concept can also be used to create rectangular pulses by replacing the capacitors by an LC-network (see Fig. 2.6) and is currently used to drive the GESA device. Each stage of pulse capacitor C_P and pulse inductor L_P is charged via a decoupling impedance L_C to the same voltage V_0. Upon closing the pulse switches, the networks are arranged in a series configuration. To avoid reflections at the load, the load impedance Z_L must be equal to the series connection of the characteristic impedance of all stages Z_S according to Eq. 2.1.

$$Z_L = n \cdot \sqrt{\frac{L'_P}{C'_P}} \tag{2.1}$$

As result, the load voltage $V_n = n \cdot V_0/2$ across the load impedance is only half of the sum charging voltage of all stages.
For short pulse lengths, the LC-network can be replaced by a coaxial cable, whereas for longer pulse lengths in the tens of microsecond range the physical dimensions of a cable become unpractical and the networks are implemented using discrete elements. Matching dynamic loads, such as the GESA device, directly to the generator is not possible [20]. Instead, the device is connected in parallel to a matched load impedance. An influence of the GESA impedance on the pulse shape can only be neglected if the current through the load impedance significantly exceeds the current drawn by the GESA device.

In the current design, the value of the load impedance is around 40 Ω, corresponding to a load current of 3 kA as compared to the nominal current of the device of 250 A.

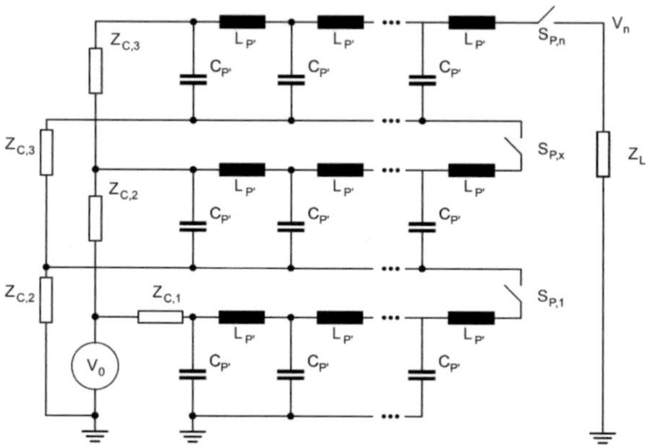

Figure 2.6: A rectangular pulse can be created using LC-chains instead of capacitors in the Marx generator.

Historically, spark gaps have been used as pulse switches for both circuits, triggered either by exceeding their standoff voltage or by using an external trigger circuit [15]. Usually only a few spark gaps are triggered and, consecutively, all remaining spark gaps follow due to an overvoltage caused by ground stray capacitances [15]. As a result, the generator can only be erected completely and cannot be switched off without external means before the voltage across the spark gaps falls below their arcing voltage, i.e. before the pulse capacitors are nearly completely discharged. Whereas spark gaps are still economical for large scale industry applications [21] due to their easy manufacturing, high-current capability and robustness, many researchers work on replacing them by semiconductor-based switching elements. Thereby, trigger delay, jitter, and electrode wear associated with certain spark gap designs can be omitted. Semiconductor switches with current interruption capability furthermore offer the possibility of active pulse shaping [22]. The simplified circuit of the generator investigated in this work is shown in Fig. 2.7. All pulse capacitors C are charged in parallel comparable to Fig. 2.5 (charging path not shown). The pulse switch consists of an active semiconductor switch T and a free-wheeling diode D as passive switch.

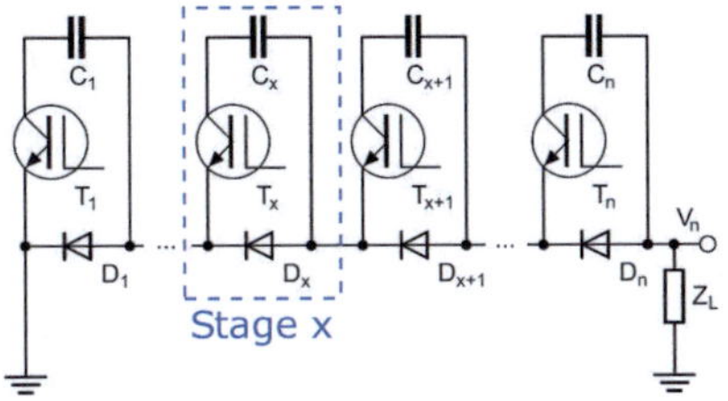

Figure 2.7: The semiconductor-based Marx generator topology, each of the n stages has a pulse capacitor C, a semiconductor switching element T and a free-wheeling diode D (charging path not shown).

In contrast to the spark-gap based Marx generator, the output voltage $V_n(t)$ can now be modulated during the pulse by changing the number of active switching elements.

$$V_n(t) = -\sum_{i=1}^{n} V_{C,i}(t) \cdot T_i(t) \tag{2.2}$$

By assigning each switching element T_i a switching function $T_i(t)$ (T_i = 1: switching element T_i closed; T_i = 0: switching element T_i open), $V_n(t)$ is given by Eq. 2.2 using the present capacitor voltage of stage i, $V_{C,i}(t)$. During an inactive period of stage i, the pulse current will by-pass the inactive stage via the free-wheeling diode D_i.
Other researchers [24] use active semiconductor switches anti-parallel to the free-wheeling diodes. After the end of the pulse, these switches are activated for a fast discharging of the load capacitance. As only a fast rising edge is important in this application, these additional switching elements have been omitted in this design.

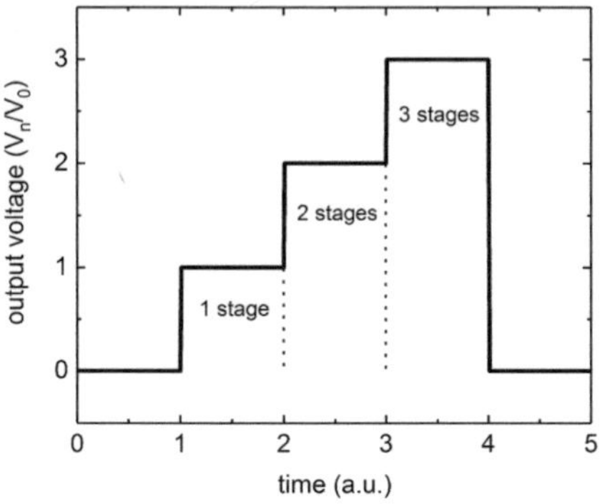

Figure 2.8: The output voltage V_n of the generator can be modulated in steps of the capacitor charging voltage V_0.

Fig. 2.8 shows an example of a step-wise arbitrary output voltage waveform V_n in steps of the capacitor charging voltage of one stage V_0, when one, two and three pulse switches are turned on with some time delay.
When operated in combination with the GESA device, it may be necessary to use the generator to control not only the output voltage (i.e. cathode-anode voltage), but also the cathode-grid voltage. Due to the modular nature of the presented topology, cathode, anode, and grid can be connected to it as depicted in Fig. 2.9: thereby, one part of the generator controls the cathode-grid voltage and one part of the generator controls the grid-anode voltage. Simultaneously deactivating stages in the upper part and activating stages in the lower part, the acceleration voltage can be kept constant while the extraction voltage is reduced. Since the grid current is below 10 % of the beam current, the unequal discharging of both parts will not be significant.

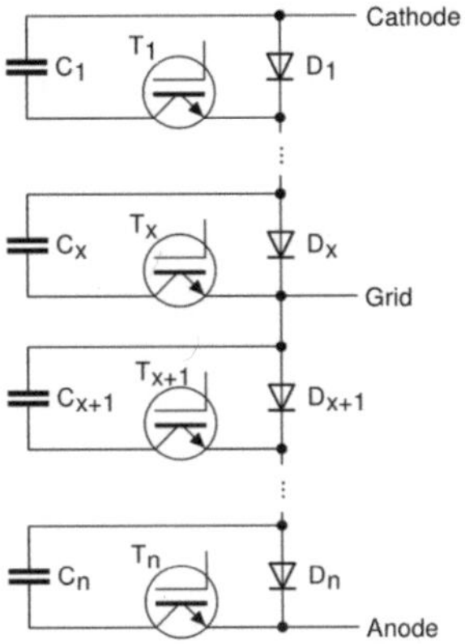

Figure 2.9: By adding a grid connection, one part of the generator can be used to control the cathode-grid voltage, whereas the other part of the generator can be used to control the grid-anode voltage.

2.2.2 Increasing the energy efficiency of the process

As discussed in the preceding section, the GESA currently is connected in parallel to the matched load of the LC-chain Marx generator. For achieving the nominal output voltage, it uses four stages charged to 60 kV with a total capacitance of 12 µF and is discharged completely during each pulse. The resulting total stored energy is given in table 2.1.

Table 2.1: Stored energy and energy loss comparison of both generator types.

Generator	Stored energy	Energy loss per pulse
LC-chain Marx generator	21 kJ	21 kJ
Semiconductor-based Marx generator	15 kJ	1.5 kJ

A comparison with the energy required by the device of 1.5 kJ (120 kV, 250 A for 50 µs) yields an efficiency of the process of close to 7 %. Since the new generator does not discharge completely in the course of the pulse, only the pulse energy needs to be resupplied. The conduction losses of approximately

3 J (1 Ω internal resistance, 250 A for 50 µs, see Sec. 6.2.2) can be neglected. Thereby, this technology offers the potential for huge energy savings by avoiding the need for a matched load.
The calculation neglects the supply voltage necessary to operate both generators, which are in the same order of magnitude. Additionally, for a highly efficient operation, the generator needs to be operated continuously without complete discharging. In the GESA experiment, the sample has to be changed after a few shots and, therefore, the generator has to be discharged for safety reasons. In future industry applications, this limitation can be avoided by automated sample manipulation.

2.2.3 Comparison to state-of-the-art pulse generators

Semiconductor-based pulsed power sources are currently of great research interest within the pulsed power community. Due to the complexity of their design, however, each power source is build according to the specific requirements of the respective application. The special challenge of building a pulsed power supply with a flexible output voltage of up to 120 kV, a maximum of 600 A pulsed current at durations of up to 100 µs with a fast rise time of around 100 ns has not been addressed before. There are several power supplies exceeding the mentioned specification in one parameter or another, but combining them in one generator is demonstrated in this work for the first time.
A much investigated approach for fast rise time, high-current, high-voltage pulses is the so-called Linear Transformer Driver (LTD) technology. The main idea behind the technology is to distribute the primary winding of a transformer on several boards, each driving a current through a single loop. The modules can be triggered simultaneously or arbitrarily to achieve a flexible output waveform. Prof. Jiang et al. [25] heavily investigated this technology and could successfully demonstrate a stack of 30 stages with 29 kV output voltage, 240 A load current and a rise time below 80 ns. While being scalable to some extend, the design relies on magnetic cores that can pose a major cost factor. The compact generator presented in [25] could only deliver a pulse of up to 170 ns in length before showing saturation effects of the core. Upscaling of the pulse length by a factor of 600 while also doubling the output current is not an economic solution.

Creating high-voltage, high-current generators with pulse lengths in the microsecond range, many companies developed interesting circuits based on the Marx generator topology investigated in this work. A very high-voltage version was presented by Applied Energetic [26], being able to charge a capacitive load up to 700 kV in several microseconds. The output waveform could be modulated by individual switching of each module. As the designed load was capacitive, the nominal charging current of 20 A lacks behind the requirement for the GESA application.

Semiconductor-based Marx generators have been investigated as replacement for thyratron-driven pulse circuits for microwave sources. A high-voltage generator under development by Diversified Technologies [27] was designed for a total output voltage of 100...500 kV and a pulse current of up to 500 A for a duration of up to 5 µs. The design could be validated up to an output voltage of 40 kV and a rise time of 300 ns was achieved. Individual control of all stages enables the possibility of active pulse shaping [14], but the high stage voltage of around 12 kV allows only for a coarse voltage adjustment. The lack in rise time and the high module voltage render this setup uninteresting for this project. Stangenes Industries [22] presented a generator able to deliver pulses of 50 kV at 250 A pulsed current for a duration of 5 µs. Again, the design is scalable, but since all parameters are inadequate for our application, a complete new design would be required. The achieved output rise rate is limited to 150 ns.

A very promising approach is currently investigated by Redondo et al. [28] as driving source for kicker magnets: Employing many paralleled very fast SiC MOSFET switches, a semiconductor-based Marx generator is under development with a module voltage of 800 V for a total output voltage of 40 kV and 3.2 kA load current. The current design is for a pulse length up to 3 µs, but the current is well above the requirements. Therefore, the stage capacitance is in the required range for this project. Until present, a prototype with an output voltage of 3.2 kV at a load current of 2.7 kA was demonstrated for a pulse duration of 2 µs. The rise time was measured to be 50 ns. The scalability of this design remains object to further studies, since a generator constructed after the prototype will be in the 4 m range for 40 kV already. In combination with the high price of the employed switching elements, this design is not suitable for investigated application.

2.3 Generator component requirements

In the course of the development of the generator, many challenges have be to overcome. The following sections define the requirements for the employed pulse components (switches and capacitors) and their control logic. Subsequently, an overview over the expected fast-varying electro-magnetic fields the circuitry has to withstand and the spacial constraints the generator has to meet is given.

2.3.1 Pulse switches and parasitic properties

As mentioned in Sec. 2.1.2, the very fast voltage rise rate across the GESA cathode of 100 kV/100 ns is one of the most challenging requirements to meet. The output voltage rise time can be expressed as geometric addition of the rise time contribution of the sub-components [29]. In first approximation, the generator can be considered a voltage source with limited rise time t_G connected to an RLC circuit. As worst-case simplification, all parasitic values of the circuit can be combined in lumped elements as depicted in Fig. 2.10. The resistive component includes all losses of the circuit, such as ohmic losses and radiated losses. L_P combines the parasitic inductive properties of the generator itself, the connection of the generator to the GESA and the GESA device.

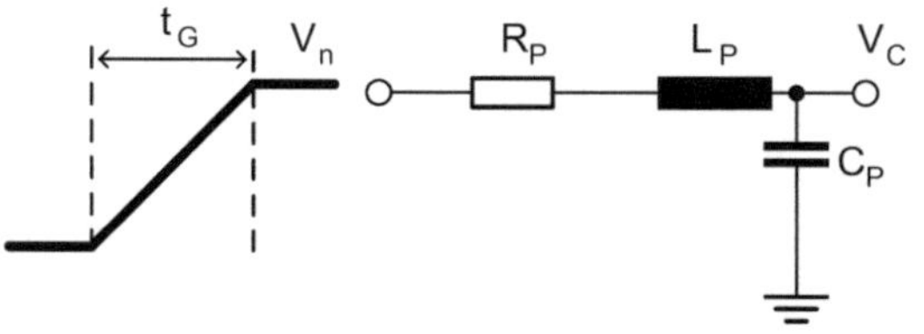

Figure 2.10: The equivalent circuit of generator and connected GESA device: the generator output voltage has a limited rise time t_G and is connected to an RLC circuit with lumped elements containing all parasitic resistive components of the circuit R_P and all inductive components (generator, connection and GESA device), charging all parasitic capacitive components of the circuit (generator, connection and GESA cathode).

As discussed in Sec. 2.1.3, the GESA cathode has a substantial capacitive component that needs to be charged. For the worst-case estimation, all stray capacitances to ground are added to the cathode capacitance to form one component C_P, including the stray capacitance of the generator and the capacitance of the connecting leads. The presented equivalent circuit is only valid as long as the cathode plasma has not ignited. Afterwards, a substantial current flows across the cathode-anode gap, shunting the cathode capacitance. Therefore, this picture is only valid to estimate the slowest voltage rise rate that can be expected with the given circuit parameters.
As first step in the design process, fast enough switching elements have to be selected as not to hinder the cathode charging process. Since the circuit parameters are not known a-priori, a target speed has to be defined for the development of the switching circuitry.
Each stage has an inherent rise time t_S. Due to many individual stages, an additional timing jitter t_J has to be assumed. A worst-case estimation for the rise time of the idealized generator t_G is represented by the sum of both rise times as in Eq. 2.3.

$$t_G = t_S + t_J \tag{2.3}$$

The precise rise time of the connected RLC network t_{RLC} will be dependent on the generator layout and physical dimensions. However, the cathode voltage rise time t_C can be calculated by the geometric sum of the components following Eq. 2.4 (compare Sec. 2.5). The resulting simulations to meet the requirements of the GESA device are presented in Sec. 6.1.4.

$$t_C = \sqrt{t_G^2 + t_{RLC}^2} \tag{2.4}$$

A target rise time t_S of 50 ns at 600 A load current in hard switching conditions was chosen for a single stage in order to allow for a sufficient safety margin in the design process. As the current will ramp up slowly in an RLC circuit, the expected rise time in the application will be much faster. Thereby, the generator output voltage rise time is mainly dependent on the circuit parameters (resistance, stray capacitance and stray inductance) instead of the switching elements. Unfortunately, commercially available semiconductor switching elements meeting these specifications are very expensive. Therefore, Sec. 3 presents a novel approach to speed up cost-efficient voltage-controlled power semiconductor devices to meet the specifications.

Using fast pulse switches also has an influence on the choice of free-wheeling diodes. The steep voltage rise across them will lead to a capacitor short during the reverse recovery time of p-n diodes, possibly destroying the device. To overcome this limitation, using fast SiC Schottky-diodes as free-wheeling diodes is imperative when employing fast pulse switches.
Resulting from the requirement of an adjustability of the output voltage in steps of 1 % of the nominal 120 kV output voltage, the maximum permissible stage voltage is around 1 kV. For avoiding failure of the employed devices, a sufficient safety margin from the maximum expected voltage across the device to its specified maximum blocking voltage has to be chosen. Increasing the voltage rating of the device, however, increases the system costs both directly, since the devices are more expensive and indirectly, since higher voltage ratings require thicker substrates with increased losses [30]. Manufacturers of power semiconductor devices intended for use in distributed supply networks recommend a safety factor up to 2.5 (device blocking voltage / expected line voltage) [30]. Due to the schematic of the generator investigated in this work (see Sec. 2.2), the voltage across the switching device is clamped by the stage capacitor in close vicinity. The manufacturer's recommendation for this DC-link-like architectures still includes a safety-factor of around 50 % due to the absence of larger network disturbances such as lightnings [30]. For a bus voltage of 1 kV, the recommendation is to use devices rated for 1.7 kV. One major effect related to the voltage de-rating is the avalanche breakdown initiated by cosmic radiation [30]. The chance of an avalanche breakdown is strongly dependent on the time integral of electric field inside the device and grows exponentially with increased voltage close to the rated voltage [30]. Felgemacher et al. could show an increased failure rate for 1.2 kV Si-IGBTs operated at 1 kV DC voltage [31]. However, since the generator presented in this work is only powered during experiments and is located under a solid concrete ceiling, both exposure time to DC voltage and cosmic ray intensity are low. Out of economical considerations, for a bus voltage of 1 kV devices with a maximum blocking voltage of 1.2 kV were used. This dimensioning of power semiconductors is commonly used in literature for pulsed power applications ([32], [33], [34], [35]). Therefore, 120 stages are required to achieve the nominal output voltage of 120 kV. If an active grid control with a 30 kV voltage swing is to be implemented, additional stages will be required. As result, the total amount of stages will be in the range of 150.

2.3.2 Pulse switch control

In multi-stage semiconductor-based pulse generators, the most efficient method of pulse switch control depends on the application. If all switches are to be switched simultaneously, a direct inductive charging of each gate is the option with the least amount of auxiliary circuitry required [36]. On the other hand, an overcurrent protection or safety relevant functions can only be implemented into the control unit. Dedicated drivers for the power semiconductors are to be used to achieve fast switching speeds; the control signal may be coupled magnetically to the stage and processed by logic on the stage [37]. The same transformer can be used to transfer both the trigger signal and the supply current [37]. However, magnetic coupling also comes with certain drawbacks: the employed transformers need to isolate the full output voltage (i.e. 120 kV) and increase the stray capacitance to ground. Secondly, no data return channel is available for detecting faults on a stage during the pulse. Most importantly, in order to drive all stages individually, many individual transformers have to be fabricated.

Instead of magnetic coupling, the presented generator uses optical signal transmission for galvanically isolated duplex communication with each stage. Details on the transmission circuit are given in Sec. 4.3.1. Thereby, stray capacitances to ground can be minimized [22].

Since every stage has to perform an individual switching sequence in the course of a pulse, there are basically two means of switching signal distribution. The first being a localized control circuitry on ground potential, controlling directly the ON/OFF sequence of every stage with no or little logic circuitry on each stage. The second method is for each stage to generate its own switching commands and the control unit only being used for starting and ending the pulse. Many researchers favor the first approach (e.g. [22], [25], [14], [34]) for its obvious advantages. Having one control circuitry for all stages allows for direct control over the switching process, necessary to implement a closed-loop control of the output voltage. If the stages are reduced to merely switches, all protection circuits have to be implemented in the control unit. Any error in the optical signal, either by EMI or wrong data from the control unit, can potentially lead to a destruction of the generator. The high number of optical connections from the control unit to each generator stage (around 150 in the presented design), all controlled simultaneously during the pulse and synchronized with respect to pulse start pose a great challenge for the control unit, both on the hardware and the software side. Moreover, approximately 300 optical fibers (two per stage to allow for duplex communication) outside of the generator increase the project complexity from a manufacturing point of view. In order to reduce the complexity of the control unit, a novel approach is investigated in this work: the switching signal generation during the pulse is not performed localized on the control unit, but is rather distributed on each single stage. Equipping each stage with a fast programmable logic, the switching signals can be pre-programmed before the pulse and executed once a synchronization signal has been sent. The control unit handles the optical distribution to enable communication with the stages. By introducing a novel optical bus system in Sec. 5.1.2, the optical connections from and to the control unit can be drastically reduced. Thereby, communication with the complete generator is possible using only 17 fiber optic connections. The logic on every stage incorporates safety functions such as maximum pulse duration monitoring, overcurrent protection, temperature measurement, and a pulse capacitor discharge functionality. The optical return channel allows for a generator emergency turn-off if any stage detects an error. For details on the control scheme, see Sec. 5.

The design was developed under the presumption that the GESA device will behave deterministically with respect to the required drive voltage pulse shape. However, since this hypothesis needs to be validated in future works, the generator was built with the possibility to use the stages also with a localized control circuitry in case a closed-loop control will become necessary in the future.

2.3.3 Dimensioning of pulse capacitors

To drive the GESA cathode, a long flat top pulse is required. For achieving the desired flat top quality of 1 kV for a pulse duration of at least 75 µs at maximum output current of 600 A, the capacitor droop needs to be minimized. The obvious solution is to employ very big stage capacitances $\mathrm{C_S}$. Since all stage capacitances are connected in series during the pulse, the equivalent capacitance $\mathrm{C_{eq,const.}}$ of an erected Marx generator using a constant number of m active stages can be calculated according to Eq. 2.5.

$$\mathrm{C_{eq,const.}} = \frac{\mathrm{C_S}}{\mathrm{m}} \tag{2.5}$$

The maximum pulse duration $t_{P,const.}$ after which the output voltage has dropped by ΔV_n at constant pulse current I_P is, hence, given by Eq. 2.6.

$$t_{P,const.} = \frac{C_{eq,const.} \cdot \Delta V_n}{I_P} \tag{2.6}$$

With the above formulas[1], individual stage capacitances C_S of 5.4 mF for a nominal output voltage 120 kV at 1 kV stage voltage are required. The total capacitance of such an arrangement would amount to 648 mF with a total stored energy of 324 kJ. On the one hand, using such large stage capacitances is a huge cost factor, on the other hand the spacial dimensions of these capacitances become impractical. In addition, most of the stored energy needs to be dumped after the experiment, increasing the complexity of the discharge circuit and being detrimental to the device efficiency.

The generator topology presented in this work enables the opportunity to use m stages necessary to achieve the required output voltage and to incorporate additional x spare stages in the design that can be activated upon demand, i.e. after the voltage has dropped by the stage charging voltage of 1 kV. The equivalent capacitance after the activation of i stages then is calculated following Eq. 2.7.

$$C_{eq,i} = \frac{C_S}{m+i} \qquad i = 0...x \tag{2.7}$$

The maximum achievable pulse length given the pulse parameters then calculates according to Eq. 2.8.

$$\begin{aligned} t_{P,var.} &= \frac{C_{eq,const.} \cdot \Delta V_n}{I_P} + \frac{C_{eq,1} \cdot \Delta V_n}{I_P} + \frac{C_{eq,2} \cdot \Delta V_n}{I_P} + \dots \\ &= t_{P,const.} + \sum_{i=1}^{x} \frac{C_{eq,i} \cdot \Delta V_n}{I_P} \\ &= \sum_{i=0}^{x} \frac{C_{eq,i} \cdot \Delta V_n}{I_P} \end{aligned} \tag{2.8}$$

[1] ΔV_n = 1 kV, I_P = 600 A, $t_{P,const.}$ = 75 µs, m = 120

With the same design pulse length, the stage capacitance sizes for the case of constant active stage count (const.) and varying stage count (var.) can be compared as in Eq. 2.9.

$$t_{P,const.} = t_{P,var.}$$

$$\frac{\frac{C_{S,const.}}{m} \cdot \Delta V_n}{I_P} = \sum_{i=0}^{x} \frac{\frac{C_{S,var.}}{m+i} \cdot \Delta V_n}{I_P} \tag{2.9}$$

Eq. 2.10 expresses the simplified ratio between the stage capacitance values in both cases.

$$\frac{C_{S,const.}}{C_{S,var.}} = 1 + \sum_{i=1}^{x} \frac{m}{m+i} \tag{2.10}$$

The total capacitance values $C_{total,const.}$ in case of constant stage number and $C_{total,var.}$ scale according to Eq. 2.11 for $m >> x$ as does the stage capacitance.

$$\frac{C_{total,const.}}{C_{total,var.}} = \frac{m}{m+x} \cdot \frac{C_{S,const.}}{C_{S,var.}} \approx \frac{C_{S,const.}}{C_{S,var.}}, m >> x \tag{2.11}$$

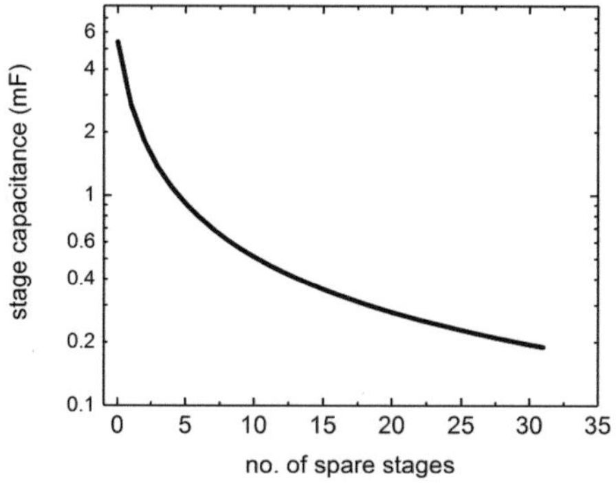

Figure 2.11: Stage capacitance value in dependence on the number of available spare stages (ΔV_n = 1 kV, I_P = 600 A, $t_{P,const.}$ = 75 µs, m = 120).

Fig. 2.11 illustrates the necessary stage capacitance value with respect to the available spare stages. Zero spare stages corresponds to the value of $C_{S,const.}$. As visible, the reduction in necessary stage capacitance is very pronounced for small numbers of spare stages but becomes less pronounced for higher numbers. Whereas more stages reduce the stage capacitance size, they increase the system costs associated with using more electronic components.
The design of this work therefore uses 29 spare stages as balance between flexibility of more stages, mechanical dimensions of the generator and stage capacitance size. The individual stage capacitances have a size of around 200 μF, and the total energy stored in the generator is equal to 15 kJ.

2.3.4 Protection against electro-magnetic interference

The great amount of high-speed logic devices involved in this project demands for adequate shielding of all components against electro-magnetic interference (EMI). A special challenge with respect to shielding in semiconductor-based Marx generators is based on the strong capacitive coupling into the control circuitry of each stage since it transiently operates at high-voltage potential during the pulse. For a low inductance, the control circuitry additionally has to be in close vicinity to the pulsed current passing through the switching elements, requiring measures against magnetic coupling. Without detailed information on parasitic oscillations that may only become measurable in the final assembly of the experiment, a rough estimation on the expected frequency content can be performed based on the expected output pulse shape.

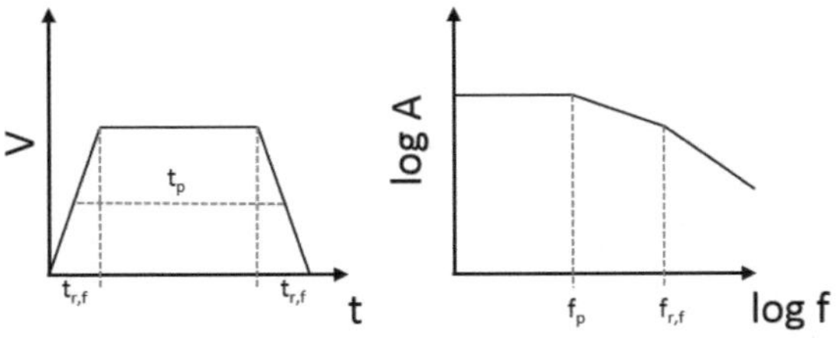

Figure 2.12: Pulse shape in time (left) and frequency domain (right). The high-frequency content of the pulse is predominantly determined by its rise and fall time. [38]

For a single trapezoidal pulse, the amplitude spectral density is determined by the pulse width t_P and the finite rise and fall time $t_{r,f}$. After the two related cut-off frequencies $f_P = 1/\pi t_P$ and $f_{r,f} = 1/\pi t_{r,f}$ the amplitude spectral density drops with 20 dB/decade and 40 dB/decade, respectively [38].

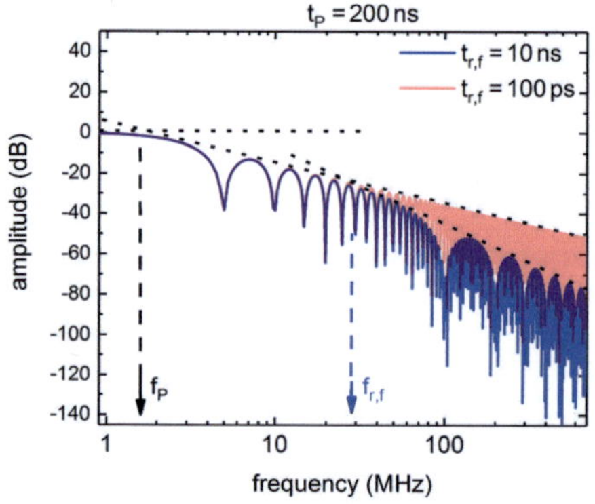

Figure 2.13: The amplitude spectral density for a pulse length of 200 ns with a rise and fall time of 100 ps (red) and 10 ns (blue). The dashed lines have a slope of –20 dB/decade and –40 dB/decade, respectively.

The assumption of a constant amplitude spectral density thereby overestimates the frequency content of the pulse as can be seen from the calculated spectrum of a trapezoidal pulse with t_P = 200 ns and varying rise and fall time as depicted in Fig. 2.13.

Whereas the generator will be build for a total rise time in the order of 100 ns, the individual switching elements switch significantly faster. As worst-case estimation, a rise time of 30 ns has be assumed based on measurements presented in Sec. 5. The corresponding cut-off frequency for this rise time is in the order of 10 MHz, being equal to a wavelength of 30 m. Compared to the physical dimensions of the generator (i.e. a few meters), the large wavelength necessitates predominantly shielding against quasi-electro- and magneto-static interference. For inserting the optical fibers, holes in the shielding enclosures are necessary. While the holes can be considered as waveguides, their cut-off frequency is much higher (for diameters in the centimeter range) than the frequency content emitted from the generator. To ensure proper damping of the interference signal, only the fraction between length L and diameter D of the opening is important. According to [39], a signal with a frequency below the cut-off frequency of an opening is attenuated by S (in dB) according following Eq. 2.12.

$$S = 32\frac{L}{D} \tag{2.12}$$

Typically, an attenuation of 100 dB is considered largely sufficient for shielding enclosures [40]. Therefore, a length to diameter ratio of greater than four was chosen for all holes throughout this project, so their influence can be neglected. At low frequencies, shielding compartments made from highly conductive material such as copper or aluminum are more effective than low-conductive, high-permeability materials such as iron [38]. Using the maximum pulse duration of 100 µs of the generator, f_P equals to 3.2 kHz. Non-magnetic materials such as copper and aluminum weaken an external non-static magnetic field purely by the counter-action of induced currents whereas in magnetic materials, such as iron, the effect is magnified by their permeability when the skin-depth is smaller than the material thickness [38]. The skin-depth of a material can be calculated according to Eq. 2.13 from the exciting frequency f, the material's permeability μ and resistivity ρ – table 2.13 gives an overview over the skin depth for the relevant materials in this work.

$$\delta = \sqrt{\frac{\rho}{\pi \cdot f \cdot \mu}} \tag{2.13}$$

Table 2.2: Skin depth for different commonly used materials.

Material	Resistivity	Relative permeability	Skin-depth (at 3.2 kHz)
Cu	$1.8 \cdot 10^{-8}\,\Omega\,m$ [41]	1	1.2 mm
Al	$2.8 \cdot 10^{-8}\,\Omega\,m$ [41]	1	1.5 mm
Fe	$5.5 \cdot 10^{-7}\,\Omega\,m$ [41]	ca. 1000 [41]	200 µm

Unfortunately, the limited resistivity of magnetic materials is detrimental to their shielding effectiveness. Additionally, saturation effects have to be taken into account when using magnetic materials. Therefore, shielding based on a combination of different materials may be advisable, with a non-magnetic layer on the outside reducing the magnetic field and protecting a inner magnetic (saturable) layer. In the following work, copper and aluminum shielding enclosures have been used, both due to their good shielding qualities (over a wide frequency range, independent of field strength) and ease of manufacturing with wall thicknesses of at least 1 mm and 2 mm, respectively. Sensitive parts, such as microprocessors and CPLDs, have been protected using an inner tinplate cabinet as second shielding (see Sec. 4.3.3).

2.3.5 Physical and mechanical constraints

Besides the electrical constraints, also mechanical constraints have to be considered. The final location of the generator will be inside an experimenting area with little ground space but a lot of headroom. The ground area for the generator is approximately 1.8 m x 1.8 m. The output of the generator will be connected to the topmost part of the GESA cathode, being at a height of approximately 2.2 m. Building the generator output to be as close as possible to this height will reduce the additional inductance introduced by wiring. For measurements, the generator should be quickly movable from the GESA experimenting site to a dedicated generator test stand located in the same building. To this end, the whole generator is designed with crane attachment points for easy movement within the building.
Being a prototype generator, easy maintenance is a major requirement. Therefore, oil insulation is not possible. All circuit boards must be quickly accessible for measurements, hence all connections are realized by plugs.
Due to the limited budget of the project, all materials used must be machinable within the institute's workshop – prohibiting the use of high-strength materials such as fiber-reinforced plastic or ceramics.

2.4 Charging and supply voltage distribution

Distribution of the pulse capacitor charging voltage is very simple in the classical Marx generator arrangement. Using active switches, new possibilities are enabled in a semiconductor-based Marx generator for transiently isolating one stage from the next. Since the circuit boards need to be powered, also the auxiliary supply voltage needs to be distributed in a similar fashion. The following section compares approaches found in literature to the most promising solution implemented in this work.

2.4.1 Pulse capacitor charging

In a Marx configuration, the charging path from one stage to the next for the pulse capacitors has to be isolated transiently during the pulse in order to avoid a short circuit [15]. Charging each stage using a transformer avoids this issue effectively [26]. On the other hand, a transformer isolating the total output voltage of 120 kV is required. Beside a strong increase in stray capacitance to ground and being a substantial cost factor, this counter-acts the idea of a Marx circuit: each component only has to withstand the stage charging voltage. Therefore, one stage is usually decoupled from the next with a high-impedance isolation element (compare Fig. 2.5). The permissible current via the isolation element thereby is dependent on the application, but should not cause a significant voltage droop of the capacitors in the course of a pulse. Under the assumption of a maximum leakage current of 10 % of the nominal pulse current of 600 A, the 1 kV potential difference from stage-to-stage necessitates an impedance of at least $R_C = 16\,\Omega$ between the stages. Using charging resistors as decoupling elements Z_C, the Marx generator as depicted in Fig. 2.5 resembles a series connection of a stage capacitor C_S and a resistor $R_C = 2Z_C$. The charging time τ_{RC} of such an arrangement consisting of n stages can be approximated by Eq. 2.14 [15].

$$\tau_{RC} = R_C \cdot C_S \cdot n^2 \tag{2.14}$$

Taking the high number of around 149 stages and the high stage capacitance of 200 µF of the design into account, the charging time can be approximated to 72 s. In reality this time is even longer since the formula assumes charging with a voltage source. The charging current in the first instant then amounts to 62 A, necessitating an expensive power supply. Charging the arrangement with a less powerful voltage source would, in turn, also increase the required charging time.

Instead of charging resistors, charging inductors are used in commercial spark-gap based Marx generators [15]. This concept has also been demonstrated successfully in semiconductor-based devices [37]. In first approximation, the current will start to increase linearly through the inductor during the pulse. To allow for the same amount of charge loss during the pulse of duration t_P as compared to the previous case, a parasitic current I_P of twice the value as for the charging resistors can be permitted (120 A). The necessary charging inductance L_C can be calculated according to Eq. 2.15.

$$L_C = \frac{V_0 \cdot t_P}{I_p} \tag{2.15}$$

With a stage voltage of V_0 = 1 kV, L_C calculates to 0.8 mH. As an air-cored inductor is required to avoid saturation effects, the physical dimensions of 149 individual inductors would become impractically large.
Employing active switches is a promising way to modulate the impedance of the charging path from low impedance during charging to high impedance during pulsing. Semiconductor switches have been demonstrated to work for several applications (see [42], [35]). The switching element used in this position has to block the current in both directions. IGBTs are in principle suitable if they are not co-packaged with a free-wheeling diode. Also MOSFETs can be used if their body-diode (compare Sec. 3.1.1) is eliminated by an additional external diode. Both possibilities, IGBT and MOSFET plus diode have a forward voltage drop in the order of 1...2 V. For generators with low stage count, this can be neglected. If 149 stages are used, however, the charging voltage of the top-most stage is reduced by a significant amount. In principle, the voltage drop per stage can be compensated using a small voltage source on each stage, powered from the supply voltage. With a charging current in the 1 A range, the required power of 1...2 W per stage, however, strains the supply voltage distribution system.
The requirement of a very low operation frequency (see Sec. 2.1.1) enables the possibility to use mechanical switches instead of semiconductors. High-voltage SMD relays are available, offering a contact resistance below 100 mΩ and can block voltage surges of up to 2.5 kV (or 1500 V RMS) [43]. They can carry a maximum current of 2 A, but even 1 A is enough to charge 149 stages to 1 kV within the required 30 s. Therefore, high-voltage SMD relays offer the most benefits of all available switches and were chosen as isolation components in the charging path of the stages. The resulting circuit is depicted in Fig. 2.14. During charging, the relay contacts $S_{1,x}$ and $S_{2,x}$ are closed, connecting the current-limited voltage source V_0 to every pulse capacitor.

While in principle the free-wheeling diodes D_x can be used to carry the return current during charging, any voltage drop across them is avoided by shorting them with relay contact $S_{2,x}$. When a pulse is executed, all relays open and the circuit is equivalent to Fig. 2.7. After the pulse, the capacitors may be charged unevenly. To prevent any equalizing current from one capacitor to the next, diodes $D_{C,x}$ are in place. Using this configuration, a negative output voltage can be generated using a positive charging voltage for the capacitors.

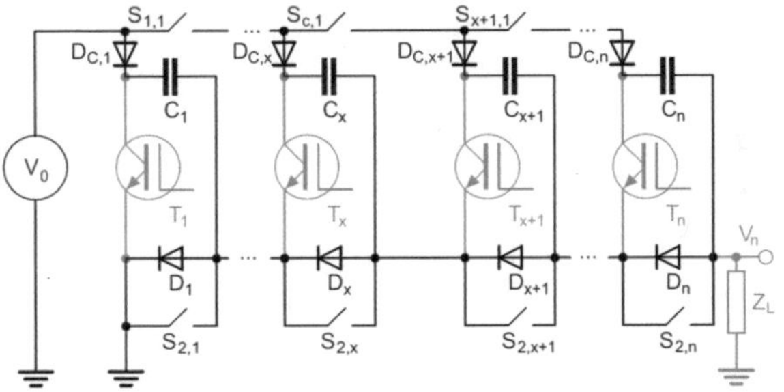

Figure 2.14: The proposed Marx circuit including the charging path, consisting of a current-limited voltage source providing a positive voltage V_0, a charging diode $D_{C,x}$, and two high-voltage relay contacts $S_{1,x}$ and $S_{2,x}$.

2.4.2 Generating the supply voltage

In addition to the power transfer to the pulse capacitors, also the logic on each stage needs to be powered and transiently isolated. A rough estimation of the required power, taking the microprocessors, the charging relays, and the optical fiber transmission (see Sec. 5.1.3) into account, the required current at 5 V level will be in the 500 mA range. For 149 stages, the required power accumulates to approximately 400 W.

As for the pulse capacitor charging voltage discussed in the previous section, distributing the voltage via a transformer is an option and has been successfully demonstrated [37]. The same arguments as in the previous section still hold true, namely the need for an expensive custom-made transformer and increased stray capacitances to ground.

Since a solution for the pulse capacitor charging voltage was already presented, one option for the supply voltage would be a step-down converter, creating the

needed supply power from the pulse capacitor voltage. To transfer the required power at a current level of 1 A (absolute maximum rating: 2 A) necessitates a charging voltage of around 400 V. Realizing a DC-DC converter with those specifications is easily possible but, in turn, demands for the pulse capacitors to be charged to nearly half the nominal voltage. Especially during the first tests of the device and trouble-shooting, the charged pulse capacitors pose a risk to the operator and necessitate additional safety measures.
Comparing the challenge of the supply voltage distribution to the challenge of pulse voltage distribution, their similarity suggest a similar solution. As mentioned before, distributing the supply power at elevated voltage levels is necessary. Since a DC-DC converter is necessary in every case, it can be designed to have a wide range of input voltages. Thereby, in contrast to the pulse voltage, a certain parasitic voltage drop across the isolation elements can be tolerated. The most efficient solution therefore is to use diodes for transient stage decoupling. Fig. 2.15 shows the circuit diagram including the current-limited voltage source providing V_S, negative with respect to ground.

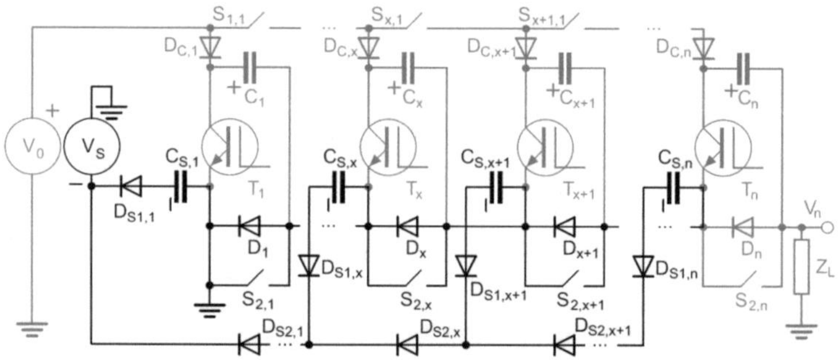

Figure 2.15: The circuit diagram of Fig. 2.14 expanded to include the negative polarity current-limited voltage source providing the supply V_S to the supply buffer capacitors $C_{S,x}$ on each stage. Diodes $D_{S2,x}$ block the negative output voltage during the pulse, whereas diodes $D_{S1,x}$ prevent equalizing currents arising from unequal discharge of the buffer capacitors.

The voltage is distributed to the stage buffer capacitors $C_{S,x}$ on all stages via a diode string consisting of n-1 diodes $D_{S2,x}$ that block the negative output voltage during the pulse. Since the buffer capacitors might discharge unsymmetrically during the pulse, diodes $D_{S1,x}$ prevent equalizing currents. As all relay contacts are closed when a pulse is not executed, the free-wheeling diodes D_x are shorted. Using a negative charging voltage has two major advantages over a positive one: primarily, the diode string $D_{S2,x}$ is wired in order to block the negative output voltage of the generator. Secondly, the supply current and the pulse capacitor charging current have opposite polarities across the relay $S_{2,x}$, canceling each other. Therefore, the limited current carrying capability of the relay of 2 A [43] is not strained by additionally conducting the supply current. The diode string $D_{S2,x}$ causes a voltage drop of around 150 V as discussed in the previous section. For transmitting the required power of 400 W at 1 A maximum current, a charging voltage of around 400 V is required. The presented design uses therefore a supply voltage of 500 V negative with respect to ground. To generate the required supply voltage for the logic circuitry, a variable input voltage DC-DC converter was designed converting an input voltage range from −300 V to −500 V into 5 V at a maximum output power of around 2.5 W. During the pulse, the relays are open and the stages are not supplied. The buffer capacitor, hence, was chosen large enough as to sustain the supply voltage for several milliseconds during the pulse.

2.4.3 Protective elements

The arrangement of both power supplies for the charging voltage and the supply voltage may subject them to fast switching transients in the course of a pulse. Fig. 2.16 (left) shows the current limited voltage source V_0 used for charging the pulse capacitors with a voltage of up to 1 kV with respect to ground, in gray its internal buffer capacitor C_B is illustrated.

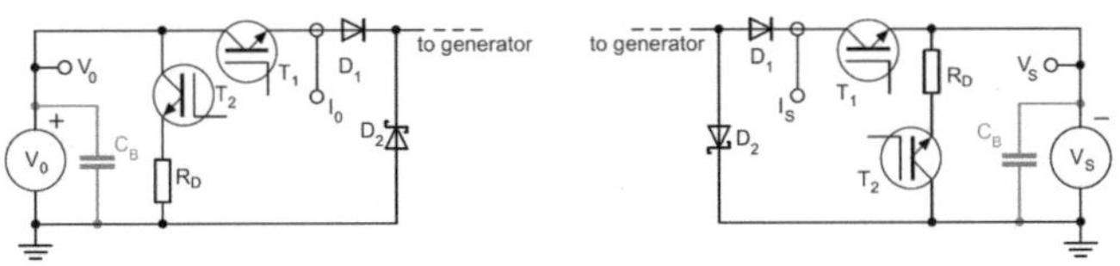

Figure 2.16: The charging power supply V_0 and the auxiliary power V_S both necessitate identical protection circuits to prevent interaction with the generator during the pulse.

The displayed protection circuitry is necessary to decouple the generator from the power supply during the pulse. It is identical for both the charging power supply V_0 and the auxiliary power supply V_S but for the polarity of all elements. Being susceptible to voltage reversal or overvoltage, diode D_2 discharges incoming transients from the generator side to ground whereas D_1 prevents an overcharging of the internal capacitance.
As mentioned in the previous section, the generator uses high-voltage relays for transient stage decoupling. One special challenge in their employment, however, is the fact that they should be operated in zero-current condition at the required voltage levels [43]. Before pulse execution, all relays have to be opened. To limit the size of the buffer capacitors on each stage, the duration between disabling the power supply, executing the pulse and (re-)enabling the power supply has to be kept at a minimum. Since the reaction time of cost-efficient power supplies is too slow for this application, T_1 is used to decouple the power supply from the generator and break the current. The current measurement I_0 (I_S, respectively) ensures zero-current condition before the pulse is executed. After the pulse, the buffer capacitors are discharged significantly. Due to the power supplies buffer capacitor C_B, a substantial equalizing current will flow as soon as the power supply is reconnected. As the relay contacts can only carry a current of 2 A, the current inrush may damage them over time. The relays therefore are protected by discharging the internal buffer capacitance C_B via T_2 and discharge resistor R_D completely during the pulse. After the pulse, when all relays are closed again, T_1 is closed and T_2 is opened, the power supply is recharging all buffer capacitors in current-limited mode. A voltage measurement of V_0 (V_S, respectively) is implemented to block reconnection of the generator in case the discharge switch fails and C_B is not discharged after the pulse. In total, four switches are required for isolating and discharging the two power supplies. As all operate simultaneously (T_1 opening before the pulse and T_2 closing; after the pulse vice versa) all switches are controlled simultaneously by magnetic coupling over one core with opposite winding directions. During pulse operation, the presented protection circuitry does not change the generator's behavior. A new challenge, however, arises when turning-off a substantial current in connection with an inductive load. This special situation is depicted in Fig. 2.17: the load current driven by the inductive load impedance Z_L commutates partly from the regular pulse current path (free-wheeling diodes D_X, blue path) to the diode string in parallel to the stages ($D_{S1,x}$ and $D_{S2,x}$, red path) and towards ground via the protection diode D as part of the protection circuit of the power supply V_S. Whereas the buffer

capacitors $C_{S,x}$ hinder a current flow in this path, the residual resistance of the generator at high currents may cause a potential drop high enough to cancel the voltage of the buffer capacitor.

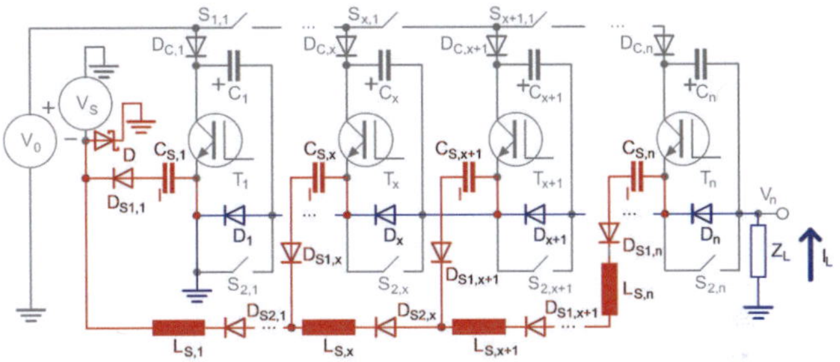

Figure 2.17: When the generator turns off, the voltage drop across the generator can lead to a partly commutation of the load current I_L into the diode string $D_{S1,x}$, $D_{S2,x}$ (red current path) in parallel to the free-wheeling diodes D_x (blue current path), driven by the inductance of the load impedance Z_L.

One possibility to prevent a parasitic current flow is to increase the impedance of the diode string by including additional inductances $L_{S,x}$. Their dimensioning can be done only after all devices have been selected and their parasitic parameters are known. From experiments shown in Sec. 6.2.2, the parasitic resistance of the free-wheeling diode path can be determined to 3 Ω. In combination with a worst-case estimation for the expected pulse current of 1.5 kA (see Sec. 5.3.4), the driving voltage across the generator can be determined to 4.5 kV. If a total driving inductance of 30 µH is assumed (see Sec. 6.1.2), the RL decay time can be calculated to 10 µs. Allowing a maximum current rise of 5 A in this time, the total inductance required in the diode string amounts to 10 mH. Divided by 149 stages, each stage needs to be equipped with a decoupling inductance of 67 µH.

2.5 Related metrology

The fast rise rate of the developed generator in combination with its high output voltage renders an accurate measurement challenging. In the following work, rise and fall time values are calculated using the 10 % to 90 % definition based

on minimum and maximum value unless noted otherwise. Rise and fall rates were calculated accordingly by dividing 80 % of the investigated quantity by the rise/fall time. Therefore, the calculated value yields a linear approximation of the interesting part of the edge only, disregarding the fringe of the edge.
For an accurate measurement of the pulse rise time, the influence of the limited bandwidth of the measurement system has to be considered. The measured signal m(t) is a convolution of the real signal r(t) and the impulse response h(t) of the connected measurement system [29] as in Eq. 2.16.

$$\mathrm{m(t)} = (\mathrm{r} * \mathrm{h})(\mathrm{t}) \tag{2.16}$$

Using the convolution theorem, the Fourier transform of the impulse response $\mathrm{H}(\omega)$ (transfer function) can be calculated by dividing the Fourier transform of the measured signal M by the Fourier transform of the real signal R according to Eq. 2.17 [44].

$$\begin{aligned} \mathcal{F}\{\mathrm{m}\} &= \mathcal{F}\{\mathrm{r}\} \cdot \mathcal{F}\{\mathrm{h}\} \\ \mathrm{M}(\omega) &= \mathrm{R}(\omega) \cdot \mathrm{H}(\omega) \\ \Rightarrow \mathrm{H}(\omega) &= \frac{\mathrm{M}(\omega)}{\mathrm{R}(\omega)} \end{aligned} \tag{2.17}$$

With a known transfer function of the measurement system, the real signal can be deduced from measurements. For most non-interacting cascaded systems [29], the total measured rise time $\mathrm{t_m}$ can be calculated from the inherent rise time of both systems $\mathrm{t_1}$ and $\mathrm{t_2}$ following Eq. 2.18 [45].

$$\mathrm{t_m} = \sqrt{\mathrm{t_1^2} + \mathrm{t_2^2}} \tag{2.18}$$

As result, the measurement system needs to be sufficiently faster than the expected signal as not to deteriorate the pulse shape. Using a measurement system with half of the rise time of the signal to be measured overestimates the rise time of the real signal by about 12 %. If the measurement system is fast enough, this error can be neglected, otherwise the measured rise time has to be corrected accordingly.

2.5.1 High-voltage measurements

The different maximum voltage amplitudes in the different stages of this project necessitate a range of suitable high-voltage dividers. An overview over the available high-voltage dividers used in this work is given in table 2.3. For the first experiments concerning the switching elements and one stage (see Sec. 3 and Sec. 4) a Philips P8932 voltage probe for up to 4 kV with 1.2 ns rise time was used. Since all presented results in these chapters are significantly slower than the specified rise time of this probe, its influence can be neglected for the experiments. In Sec. 5, measurements of up to 8 kV are presented, with a maximum rise time of around 30 ns. With a rise time of 15 ns, the employed divider (HVT-40-RCR) causes an error of 11 % for the fastest signal.

Table 2.3: Available high-voltage dividers and their specifications. [46], [47]

Type	Max. working voltage (DC)	Rise time	Bandwidth
Philips P8932	4 kV	1.2 ns	290 MHz
HILO-TEST HVT 40 RCR	40 kV	15 ns	23 MHz
HILO-TEST HVT 240 RCR	240 kV	48 ns	7.3 MHz

Only for the measurements on the full-scale generator assembly (see Sec. 6, HVT-240-RCR) the picture changes as the divider's capacitance and resistance have an influence on the measured output pulse. To ensure a correct interpretation of the measured signals the divider's transfer function was verified in the experiment to exclude measurement errors due to stray capacitances and stray inductances arising from the installation in the experiment. By using a HILO-TEST high-voltage impulse generator with a specified rise time of 3 ns a pulse of 200 ns length was measured both with a very fast high-voltage probe (P8932) and the HVT-240-RCR divider. The time-domain graph is shown in Fig. 2.18a, illustrating the limited rise time of the divider. A Fast Fourier Transformation (FFT) of both pulses yields their amplitude spectral density (see Fig. 2.18b) and, according to Eq. 2.17, the transfer function of the divider. Its corner frequency (−3 dB) at 7.7 MHz is in reasonable agreement with the datasheet value. In addition, the graph shows the frequency response of a first order low-pass filter (LPF) with the same corner frequency. The good agreement of both traces allows therefore to approximate the response of the high-voltage divider in time-domain with an RC element (see Sec. 6.2.5) and is in agreement with literature [29]. This behavior is only relevant for the fast leading edge of the measured pulses and can be neglected for other measurements.

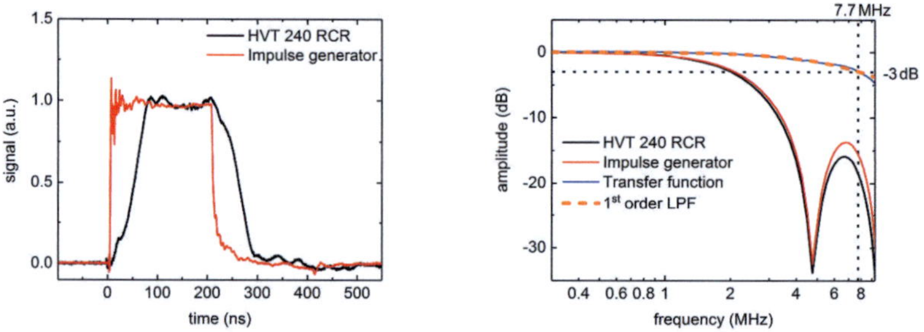

(a) Time-domain response of a high-voltage divider with limited rise time to a fast-rising signal.

(b) Normalized amplitude spectral density of both pulses.

Figure 2.18: In time-domain, the high-voltage divider responds to a step-excitation within its specified rise time. From the difference in the amplitude spectral density of both traces, the transfer function can be deduced.

2.5.2 High-current measurements

All high-current measurements in this work have roughly the same amplitude of several hundred ampere. For short pulse lengths on single boards (Sec. 3 and 4), a fast coaxial shunt resistor with a flat frequency response up to 1 GHz was used [48]. Due to its limited energy absorption capability, it was replaced for pulses longer than several microseconds with a Pearson current monitor type 110 with an usable rise time of 20 ns (Sec. 5 and 6).

2.5.3 Data acquisition

The presented time-depended data was recorded using the oscilloscopes listed in table 2.4. Prior to all measurements, the voltage probes were balanced for the respective channel input capacitance. After the measurements, the cable propagation delay of all signals was compensated with a residual delay of less than 1 ns.

Table 2.4: Available oscilloscopes and their specifications. [49], [50]

Type	Resolution	Bandwidth	Sample rate
LeCroy WaveRunner 204MXi-A	8 bit	2 GHz	10 GS/s
LeCroy HDO6104	12 bit	1 GHz	2.5 GS/s

In order to avoid measurement errors due to induced voltages, measurement loops are kept as small as possible and the cables are routed in parallel to the grounding lines. By using a single grounding point for all equipment and installing ferrites around the measurement cables (when necessary) sheath currents were minimized. The absence of important disturbances in the measurement circuitry was verified by reference measurements with open and shorted cables, while the reproducibility of each measurement was ensured by many consecutive experiments.

3 Gate-boosting investigations

The main challenge in fast switching of voltage-controlled semiconductor devices is a fast charging of the input capacitance. Whereas inherently fast switching elements, such as SiC MOSFETs, are available, they are currently very expensive with respect to their switching power. The gate-boosting circuit introduced in this chapter allows for a considerable increase in resonant frequency of the input network and, therefore, for a tremendous increase in switching speed. The presented results validate an eight-fold increase in switching speed for commercial IGBT devices, enabling their use as fast and cost-efficient switching elements in pulsed power circuits. Major results of this work presented in the following chapter have already been published as [51], [52], and [53].

3.1 Device architectures

Since the development of the first semiconductor-based switching devices in the early 1900's [54], a multitude of device architectures have been invented and are described in detail in literature. The basic concept is to modulate the conductivity of one part of the semiconductor from low conductivity (in blocking state) to high conductivity (in conducting state). From an economical point of view, only switching elements fabricated in large quantities are interesting for the use in a generator employing at least 149 individual switches. Additionally, the elements must be able to interrupt high currents (maximum 600 A) and block voltages in the order of 1 kV at high switching speeds (approximately 50 ns, see Sec. 2.3.1). At present, the only competitive devices, namely the Metal-Oxid-Semiconductor-Field-Effect-Transistor (MOSFET) and the Insulated-Gate-Bipolar-Transistor (IGBT), are investigated in this work. In addition to semiconductor devices fabricated from silicon (Si), new high band gap materials have become available over the past years. The larger band gap allows for a higher internal critical electrical field and, hence, smaller structures can be used to block the same magnitude of voltage. Thereby, both switching speed (due to smaller capacitances and higher electron drift velocities) and conduction losses (due to shorter conduction paths) could be improved [55]. Especially silicon carbide (SiC) has recently received a lot attention from the industry as manufacturers could mature the fabrication processes and supply stable devices with interesting performance. On account of a band gap with thrice the value of Si[1], a ten-fold increase of internal electric field can be achieved [55]. Thereby, thinner devices can be used with a 300-fold smaller on-state resistance. In addition, the higher electron drift velocity increases the switching speed of those devices. However, the higher band gap also increases the p-n junction voltage to around 2.7 V, rendering the material only interesting for unipolar devices such as MOSFETs. Other technologies such as gallium-nitride (GaN) may offer even superior performance, but are currently not technologically mature enough to be available in large quantities [57].

3.1.1 MOSFET

The basic MOSFET architecture is shown in Fig. 3.1a. The electric field between drain and source terminal is blocked by the weakly doped n-drift region, as there is no conductive path between the n+ region in contact with the source terminal. The gate terminal, isolated by a thin oxide layer, serves as control terminal for the device. By applying a positive voltage with respect to the source terminal, the Fermi-level in the p-doped material close to the gate contact can be raised. If the voltage exceeds the threshold voltage $V_{GS,Thres.}$, electrons start to populate the formerly empty conduction band (charge carrier inversion), forming a conductive channel between drain and source for the load current I_L. As the current (in the p-region) is carried by minority carriers, the transition from conducting to blocking state (if the gate voltage falls below the threshold voltage) does not involve bipolar recombination and, hence, is very fast. Since no p-n junction is involved, the conduction losses of the MOSFET are determined purely by its resistive properties.

[1] Si: 1.12 eV; SiC (4H-SiC): 3.2 eV [56]

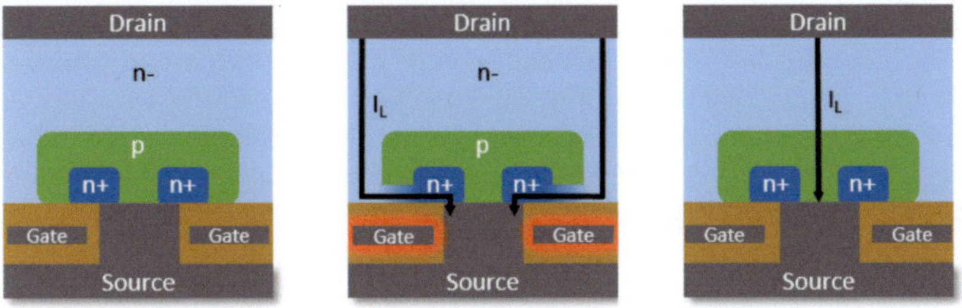

(a) The MOSFET architecture with drain, gate, and source terminal without applied voltage.

(b) By charging the gate terminal positively with respect to the source terminal, an inversion layer is formed in the lower part of the p-doped region, enabling electron flow from source to drain.

(c) Under reverse bias (drain negative with respect to source terminal), the inherent (body-) diode in the MOSFET structure becomes conductive.

Figure 3.1: The MOSFET architecture under blocking conditions (a), forward conducting conditions (b) and reverse conducting conditions (c).

Especially at high blocking voltages, the limited critical field strength of silicon leads to very thick structures causing considerable on-state resistance. Therefore, the MOSFET architecture at blocking voltages in the 1 kV range necessitates paralleling many devices. Specialized RF MOSFETs with considerable blocking voltage and rise time are available[2] but come at device costs of around 35€ (equal to 4.1 kW/€[3]) per device and require dedicated drivers [60] at around the same price. As mentioned before, SiC MOSFETs have much lower conduction losses per device size[4] and therefore are much cheaper (17€, equal to 5.6 kW/€ [59]) with respect to switched power. Nevertheless, their limited maximum current capability still requires a substantial amount of parallel devices (compare design criteria Sec. 4.1.1).
Another feature of the MOSFET architecture is its diode-like behavior under

[2] i.e. IXYS DE475-102N21A: $V_{DS,max.}$= 1 kV, $I_{DS,pulsed}$= 144 A, t_r= 5 ns, $R_{DS,on}$= 450 mΩ [58], 35€ [59]

[3] Power calculated as max. voltage times max. pulsed current.

[4] i.e. C2M0080120D: $V_{DS,max.}$= 1.2 kV, $I_{DS,pulsed}$= 80 A, t_r= 20 ns, $R_{DS,on}$= 80 mΩ [61]

reverse bias conditions. When the drain-source channel is biased negatively (Fig. 3.1c), the intrinsic p-n junction between drain and source becomes conductive. Being a parasitic part of the architecture, this so-called body-diode can not be altered independently of the switching architecture. Therefore, its performance is often inferior to specialized switching diodes, giving rise to switching losses [62]. This is especially true for SiC devices, as bipolar high band-gap devices suffer from high p-n voltages and slow reverse recovery times. A common way to overcome this limitation is the use of fast SiC Schottky diodes both in series (eliminating the body-diode) and anti-parallel (replacing the body-diode) to the device and the diode [63].

3.1.2 IGBT

The IGBT architecture is widely used in high-power applications with moderate switching speeds of several tens of kHz [56]. Its equivalent circuit can be seen in Fig. 3.2a.

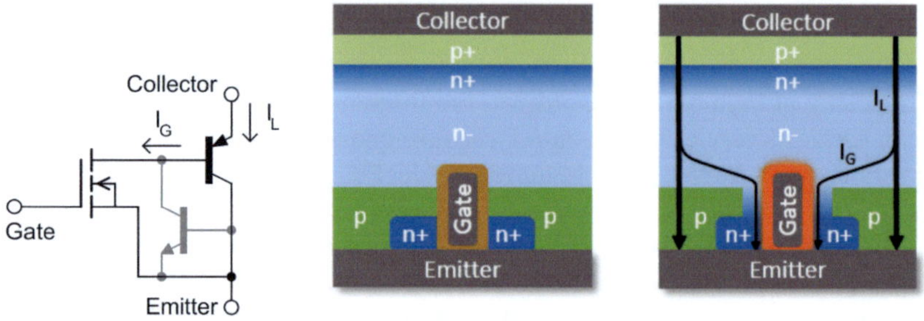

(a) Equivalent circuit of an IGBT, consisting of a MOSFET controlling a PNP transistor. Together with the parasitic NPN transistor, a thyristor structure is formed.

(b) The IGBT architecture with collector and emitter terminal without applied voltage.

(c) In the conducting state (gate voltage above threshold voltage), the base current for the PNP transistor is supplied by the MOSFET structure.

Figure 3.2: The IGBT equivalent circuit (a), its structure under blocking (b), and forward conducting conditions (c).

The voltage is blocked by a PNP transistor between collector and emitter terminal, also carrying the load current I_L. The drawback of a transistor, namely a low input impedance resulting a high required drive power in on-state conditions is avoided by controlling its gate current I_G via a MOSFET integrated in the device itself. Figure 3.2b illustrates the resulting layer diagram[5]. In conducting state (Fig. 3.2c), when the voltage applied on the gate contact exceeds the threshold voltage, the base current of the PNP transistor is delivered by the MOSFET structure close to the emitter contact, closing the switch. In contrast to the pure MOSFET architecture, the bipolar nature of the current in the transistor offers a much lower on-state resistance. However, due to the involved p-n junction, the voltage drop across the element at low currents is at least the saturation voltage of the junction $V_{CE,sat.}$ [62]. Another property of the bipolar current is the significant recombination time of charge carriers during device turn-off, leading to slow current decay ("tail-current"). By increasing the density of trap states and therefore the charge carrier recombination rates ("life-time killing" [55]) a high tail-current can be omitted at the expense of increased conduction losses. One inherent drawback of the IGBT architecture is the combination of a PNP-transistor and a parasitic NPN-transistor (Fig. 3.2a, gray) forming a thyristor structure. Operating an IGBT outside of the datasheet ratings may result in the ignition of this structure ("latch-up"), causing a device short and, if sufficient energy is provided, device destruction [65]. In more recent device architectures, however, this phenomenon is of lesser concern [66]. As no body-diode is present in the design, the devices can be bought packaged together with a suitable anti-parallel diode. The high-current capability per active area, that can be achieved in IGBTs, leads to very low production costs. Their prices are with around 5€ [59] (equal to 76 kW/€[6]) more than a factor of ten lower than that of MOSFETs. To increase their current capability, they can be bought pre-packed in modules consisting of several devices in one package. However, throughout this work only single devices in TO-247 packages have been used, as their prices per switched power are considerably below those of modules[7]. More importantly, paralleling individual devices allows for a short and symmetric layout of all device leads which is key for a synchronous

[5] Field-Stop Trench IGBT, for different architectures see [64]

[6] i.e. NGBT40N120IHLW: $V_{CE,max.}$= 1.2 kV, $I_{CE,pulsed}$= 320 A, $V_{CE,sat.}$= 1.9 V [67]

[7] 2018: Prices around 1 kW/€ (Infineon – BSM10GD120DN2) to 9 kW/€ (FZ400R12KE3 – Infineon) for modules, 35 kW/€ (NGTB40N120FL3WG – ON Semiconductor) to 10 kW/€ (IXYH20N120C3D1 – IXYS) for TO-247 devices (1200 V, max. peak current, [59])

turn-on of all devices at high switching speeds (compare Sec. 4.1.1).
The major benefit of MOSFETs over IGBTs is their ten-fold superior switching speed. Using the gate drive technique presented in the following section, switching speed of cost-efficient IGBTs becomes comparable to that of MOSFETs. Thereby, the costs for switching elements in future pulsed power projects can be dramatically reduced.

3.2 Fast switching of voltage-controlled power semiconductors

3.2.1 Limiting factors

For power semiconductor devices, the transition speed from low conductivity (i.e. blocking) state to high conductivity (i.e. conducting) state is a key figure of merit. In the presented project, fast switching elements are a crucial component of the circuit in order to achieve a fast-rising output voltage (compare Sec. 2.3.1).
The switching losses E_S are given by the time integral over the product of the voltage drop across the device V and the conducted current I from the start of the switching process at t_0 to the end of the switching process t_e according to:

$$E_S = \int_{t_0}^{t_e} V(t) \cdot I(t) dt \tag{3.1}$$

By minimizing the switching time, switching losses are greatly reduced. As discussed in Sec. 3.1, in both MOSFETs and IGBTs the conductivity modulation is achieved by charging the gate-emitter capacitance C_G. In addition to the capacitive component, the gate driver needs to charge, the device also has a parasitic resistive and inductive component, R_P and L_P respectively. Whereas the resistive component may be a discrete element such as an internal gate resistor, the inductive component arises mainly from the bond wires, the device leads and the circuit layout. During circuit design, the stray inductance of the gate circuit has to be minimized in order to prevent significant turn-on and turn-off delays [68]. These contributions are illustrated in Fig. 3.3. The challenge of a fast charging of the resulting second-order low-pass filter will be addressed in the following sections.

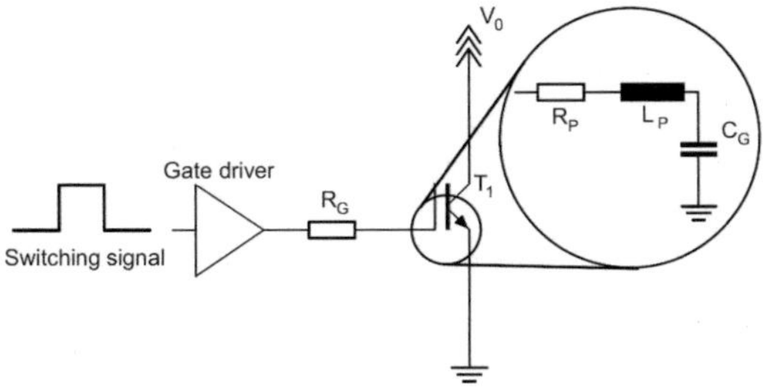

Figure 3.3: The gate lead of the voltage-controlled power semiconductor T_1 has a parasitic ohmic resistance R_P, a parasitic lead inductance L_P and terminates on the gate capacitance (C_P). Usually, the device is driven by a gate driver via a gate resistor (R_G).

Apart from the aforementioned gate-emitter capacitance, the gate-collector capacitance C_{GC} and the collector-emitter capacitance C_{CE} as depicted in Fig. 3.4a play an important role in the switching behavior. In steady-off state, C_{GC} and C_{CE} are charged to approximately the bus voltage V_0. Once the device starts to turn on, the voltage across collector and emitter collapses and the capacitances C_{GC} and C_{CE} are being discharged. The discharge current I_{GC} for discharging C_{GC}, however, has to be supplied from the gate drive circuit:

$$I_{GC} = C_{GC} \cdot \frac{dV_0}{dt} \tag{3.2}$$

This effect, named Miller-effect [64], becomes important as soon as the threshold-voltage across the gate capacitance is reached and the device turns on. At constant current charging, the voltage across the gate capacitance exhibits a plateau (Miller-plateau), thereby slowing down the gate charging process and therefore the switching speed. Fig. 3.4b illustrates another detrimental stray parameter reducing the switching speed: the parasitic emitter inductance L_E shared by the load current I_L and the gate drive current I_G. Once the device starts conducting, the steep rise in the load current via L_E shifts the emitter potential with respect to the gate drive circuit, reducing the gate drive current. Therefore, careful circuit design reducing current loops and minimizing capacitive coupling is key when employing fast switching circuitry.

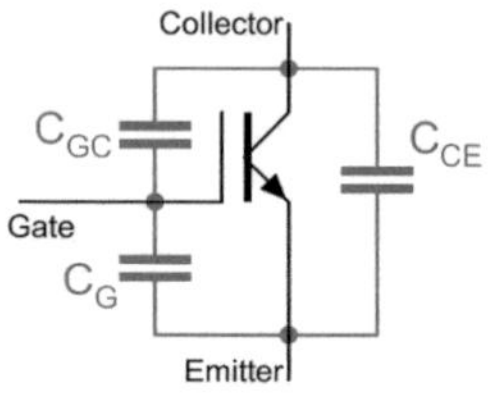

(a) In addition to the gate-emitter capacitance C_{GE}, the gate-collector capacitance C_{GC} and the collector-emitter capacitance C_{CE} determine the switching behavior of a device.

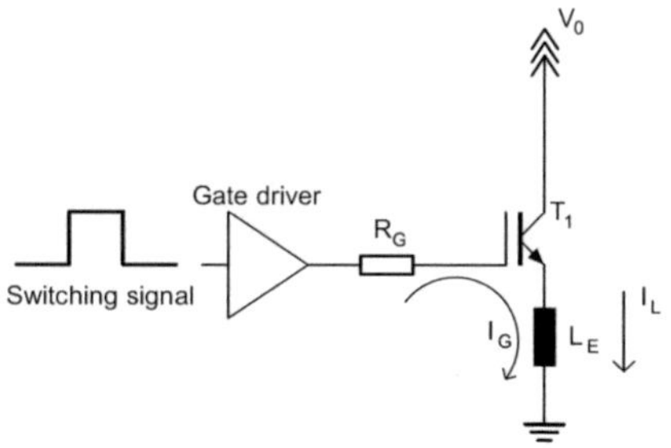

(b) On turn-on, the increasing load current I_L through the parasitic emitter inductance L_E shared with the gate drive circuit counteracts the gate charging current I_G, hindering the gate charging process.

Figure 3.4: Parasitic elements with influence on the gate charging process.

3.2.2 Conventional approaches

For addressing the challenges discussed in the previous section, many standard approaches are known. In reference to Fig. 3.3, the resonant circuit formed by the gate capacitance and the parasitic circuit inductance is one of the main factors responsible for a slow gate charging process. The resonant frequency of the circuit can be calculated according to Eq. 3.3.

$$\omega_0 = \frac{1}{\sqrt{C_G \cdot L_P}} \tag{3.3}$$

Fig. 3.5 shows a varying degree of damping introduced by selecting different gate resistors. When charging the gate to the nominal gate voltage of 15 V[8] the circuit can be adjusted to overdamped (red, solid), critically damped (black), and underdamped (blue) conditions. Since the device starts to turn on as soon as the gate threshold voltage (approximately 7 V) is reached a reduction

[8] The actual value for the respective device is listed in the datasheet

in damping is useful to increase the switching speed. One limitation to this approach is the ringing in the step-response of a very weakly damped resonant circuit. The gate voltage falling significantly below the nominal gate voltage causes increased losses during the switching process and may even destroy the switching element by thermal stress [69]. As an overvoltage at the gate may cause damage to the device [70], excessive voltage overshoot has to be prevented. Whereas the approach to use a higher-than-specified gate voltage (Fig. 3.5 – red, dashed) also enables a fast charging of the gate capacitance, again the device is operated outside of the recommended drive conditions.

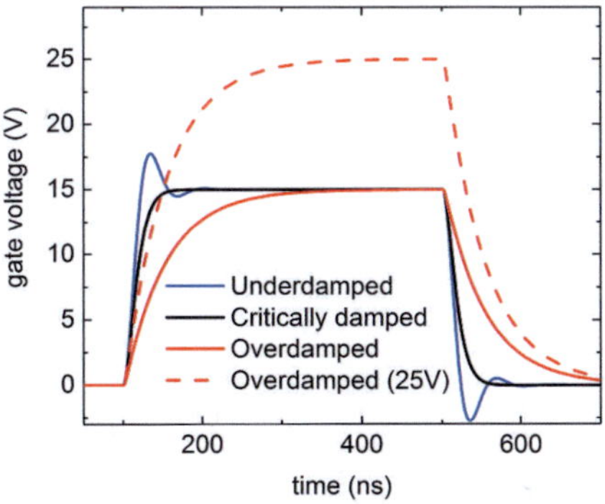

Figure 3.5: Conventionally, the gate-voltage threshold is reached faster by a reduction of the damping in the gate drive circuit (black, blue) or employing a higher gate drive voltage (red, dotted).

If the standard gate drive circuit is to be used, the damping therefore should be adjusted to critically-damped or slightly underdamped conditions, with the maximum voltage overshoot within the permissible limits of the datasheet. Thereby, the gate driver has to be able to deliver the required current. Additionally, the resonant frequency of the gate circuit can be reduced by keeping the parasitic inductance as low as possible. Commercial solutions are available where the gate drive circuit is monolithically integrated into the power semiconductor [55]. Since these solutions are limited to a specific field of application, their use in pulsed power generators is usually not cost-efficient. However, decreasing the parasitic inductance during the circuit design process is very important. Special care has to be taken in order to minimize the length of the mutual current path of gate charging current and load current. One possibility is to use special device packages featuring Kelvin-Emitters (different emitter pins for load and gate drive current), allowing a reduction of switching losses of up to a factor of 3 [71]. Whereas those packages are used with great success [72], they are only available for selected devices and at comparably high costs. Since the turn-on time is reduced with reduced load current [73], paralleling many devices is also an alternative. Consequently, the increase in switching speed with this method results in increased circuit complexity and increased costs for the additional switching elements and their drive circuitry.

3.3 Fast switching by gate-boosting

3.3.1 Analytical approach

As discussed in the previous section, the inherent drawback of the standard (i.e. resistively-coupled) gate drive circuit as shown in Fig. 3.6 is the low-pass filter consisting of the gate capacitance and the parasitic inductance. Usually, the gate drive circuit consists of a push-pull stage (M_1, M_2) charging the gate capacitance of T_1 via gate resistor R_G up to the gate drive voltage V_{GD} or discharging it completely (a negative gate voltage is discussed in Sec. 3.3.2). The maximum resonant frequency of this circuit, being determined by the inherent device parameters (i.e. gate capacitance and device lead inductance), cannot be increased above a certain value for a given device. To overcome this limitation, there exist several approaches in literature to use a gate drive voltage significantly above the rated value, while not exceeding the specified value in steady-state. Those approaches are referred to as "gate-boosting". Some researchers (i.e. [64] or [74]) use two different voltage levels during transient turn-on (+24 V) and steady-on (+15 V). Others [75] propose to discharge a capacitor charged at a voltage of +60 V into the gate and prevent an overvoltage by clamping it using a Transient-Voltage-Suppression diode.
An interesting, but very challenging solution is employing a current source as gate driver [76]. Whereas these topologies work well, they necessitate additional circuit elements and, hence, increase the circuit complexity.

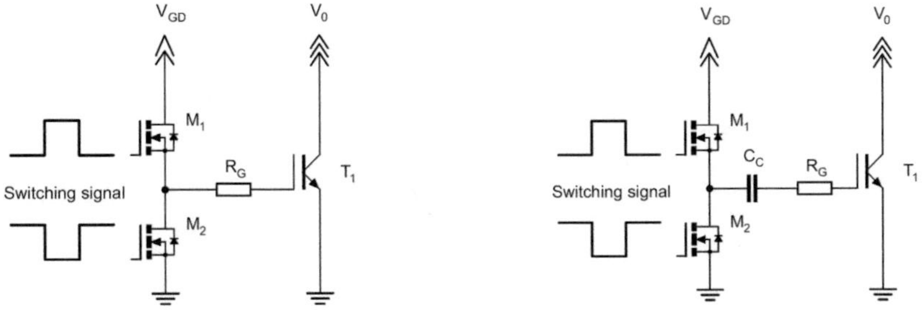

(a) The standard resistively-coupled gate drive circuit.

(b) The proposed capacitively-coupled gate drive circuit.

Figure 3.6: Simplified schematics of the discussed gate drive circuits.

As new drive method, the investigated topology in this work is a capacitively-coupled gate drive circuit. The simplified schematic is shown in Fig. 3.6b. The only additional element in the circuit is the coupling capacitor C_C in series to the gate resistor. The approach allows to use very high gate drive voltages while respecting the datasheet limits if designed correctly. For a better understanding, the equivalent circuit of Fig. 3.6b is shown in Fig. 3.7.

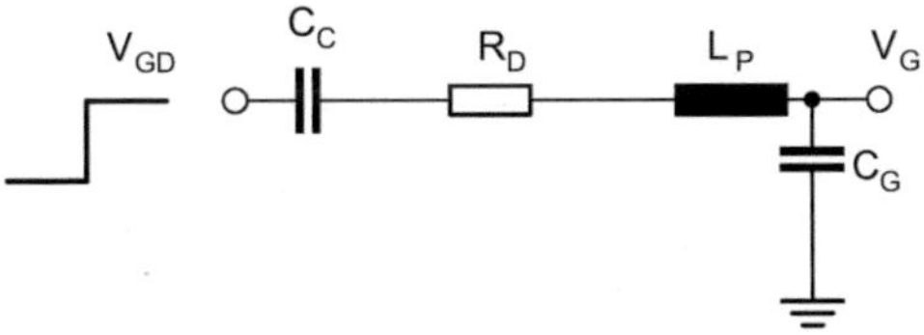

Figure 3.7: Simplified equivalent circuit of Fig. 3.6b.

The gate drive voltage V_{GD} charges the gate. All damping resistors, either discrete elements such as the gate resistor or internal resistive elements of the device, are included in the equivalent damping resistor R_D. L_P represents all parasitic inductance, including the circuit inductance, the device lead inductance and bond wire inductance. The gate voltage V_G is measured across the gate capacitance C_G. In the following calculations, the gate capacitance C_G is assumed to be constant, which holds not true under close investigation [64].

However, the non-linear behavior of the gate capacitance was accounted for using an increased apparent capacitance value as can be obtained from the gate-charge versus gate-voltage plot as given in the datasheet. The comparison of simulations with measurements as presented in Sec. 3.3.3 validates the prior simplification assumption.
The basic concept of this circuit is the combination of the coupling capacitor, the damping resistor and the gate capacitance forming a damped capacitive voltage divider. Prior to turn-on, all capacitances are discharged. Under the assumption of a sufficiently fast driver circuitry and low-value damping resistors, nearly the complete gate drive voltage is available across the parasitic inductance for achieving a high gate charge current gradient during turn-on. The coupling capacitor is chosen such, that once the gate capacitance reaches its nominal value $V_{G,nom.}$, the coupling capacitor takes up all the excess voltage. Its value can be calculated according to Eq. 3.4. Therefore, the device specification regarding the maximum gate voltage is not exceeded.

$$C_C = \frac{C_G}{\frac{V_{GD}}{V_{G,nom.}} - 1} \tag{3.4}$$

A different approach of understanding this circuit offers a glance at the transfer function G(s).

$$\frac{V_G}{V_{GD}} = G(s) = \frac{\frac{1}{C_G \cdot s}}{R_D + \frac{1}{C_C \cdot s} + L_P \cdot s + \frac{1}{C_G \cdot s}} \tag{3.5}$$

For simplification, the equivalent capacitance C is defined by the series connection of the coupling capacitance and the gate capacitance.

$$C = \left(\frac{1}{C_G} + \frac{1}{C_C} \right)^{-1} \tag{3.6}$$

Using Eq. 3.6, the transfer function can be simplified.

$$G(s) = \frac{\frac{1}{C_G \cdot s}}{R_D + \frac{1}{C \cdot s} + L_P \cdot s} = \frac{1}{R_D \cdot C_G \cdot s + \frac{C_G}{C} + L_P \cdot C_G \cdot s^2} \tag{3.7}$$

The characteristic equation of Eq. 3.7 now has the well-known shape of the RLC series resonant circuit.

$$R_D \cdot C_G \cdot s + \frac{C_G}{C} + L_P \cdot C_G \cdot s^2 =$$
$$s^2 + \frac{R_D}{L_P} \cdot s + \frac{1}{L_P \cdot C} = 0 \quad (3.8)$$

The undamped resonant frequency of this characteristic equation can be calculated according to Eq. 3.9.

$$\omega_0 = \frac{1}{\sqrt{L_P C}} \quad (3.9)$$

Effectively, introducing a coupling capacitor C_C increases the circuit's resonant frequency and thereby speeds up its transient response. The increase in speed comes at the cost of a higher required gate drive voltage. An exemplary set of gate voltage traces[9] for increasing gate drive voltage can be seen in Fig. 3.8. For each voltage step, the coupling capacitor was decreased to keep the steady-state gate voltage constant. Since the damping resistor was not adjusted, the damping decreases from overdamped to underdamped conditions.
The resulting gate voltage rise time of the traces in Fig. 3.8 are displayed in Fig. 3.9. The linear increase in gate drive voltage leads to an exponential-like decrease in the gate voltage rise time.

[9] Numerical values: $R_D = 4\,\Omega$, $V_{G,nom.} = 16\,V$, $L_P = 25\,nH$ and $C_G = 18\,nF$

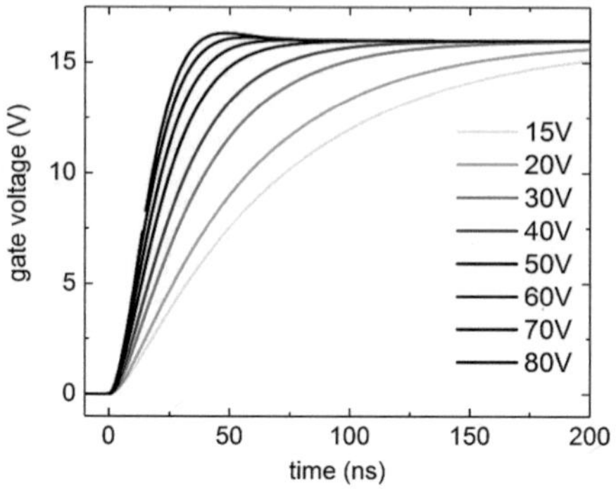

Figure 3.8: Simulated gate voltage traces for different gate drive voltages. Increasing the gate drive voltage allows for a fast charging process of the gate capacitance.

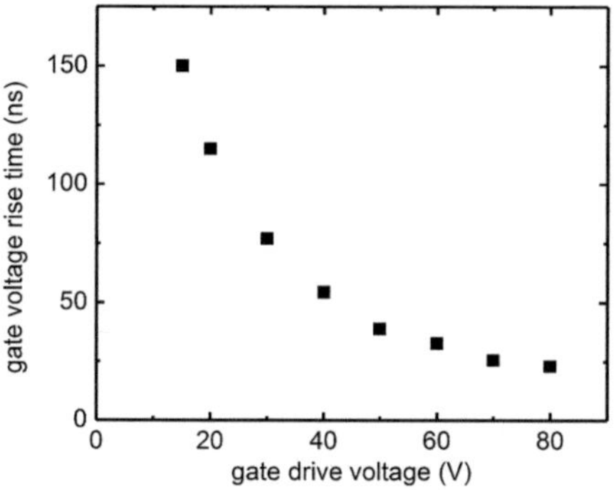

Figure 3.9: Gate voltage rise time in dependence of the gate drive voltage. By increasing the gate drive voltage, the voltage rise time across the gate capacitance is decreased significantly.

Fig. 3.10 shows a measurement (black trace) of the gate voltage on an IGBT (NGTB40N120) operated with 80 V gate drive voltage. As can be seen, the voltage overshoot at the device leads is around 30 V and therefore significantly outside of the datasheet rating (±20 V [67]) for the specific device. By simulating the circuit as depicted in Fig. 3.11 in SPICE (C_G = 18 nF, L_P = 8 nH), the measured signal can be reproduced with sufficient accuracy (red trace). The blue dashed line represents the voltage as simulated across the internal gate capacitance. As can be seen, the voltage spike occurs only across the parasitic inductance in the circuit, while the voltage across the internal gate capacitance remains within the datasheet limits. In addition to the previously discussed parameters, the circuit also incorporates an internal gate resistor R_I that is present in certain voltage-controlled power semiconductors [77]. Its value was determined by iterative adjustment to 0.8 Ω. This simulation is verified by measurements directly on the die in Sec. 3.3.3.

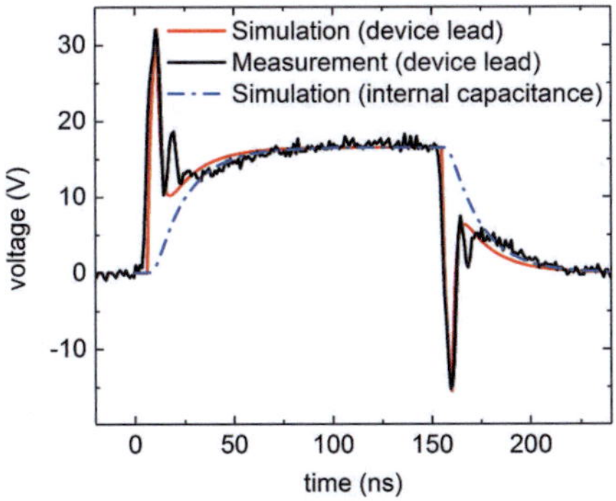

Figure 3.10: Measurement and SPICE simulation of the gate voltage when operating a device at 80 V gate drive voltage. While the voltage overshoot at the device leads is substantial, the sensitive gate capacitance is not overcharged.

3.3.2 Circuit considerations

As compared to the simplified schematic shown in Fig. 3.6b, additional elements are necessary to ensure the functionality of the circuit. The Device-Under-Test (DUT) T_1 investigated in the experiments is considered an ideal switch, whereas its parasitic components inside of the device housing, namely the gate-emitter capacitance C_G, the internal gate resistor R_I, and the parasitic inductance L_P, are drawn as discrete components. For the following experiments, the gate drive voltage V_{GD} could be adjusted between 15 V and 80 V. The MOSFET half-bridge driving the circuit has to be significantly faster than the expected gate voltage rise time at the DUT, while being able to block V_{GD} including a safety margin. The devices used for M_1 and M_2 are FDT86102LZ MOSFETs with a specified rise time of typically 1.2 ns with a maximum blocking voltage of 100 V [77].

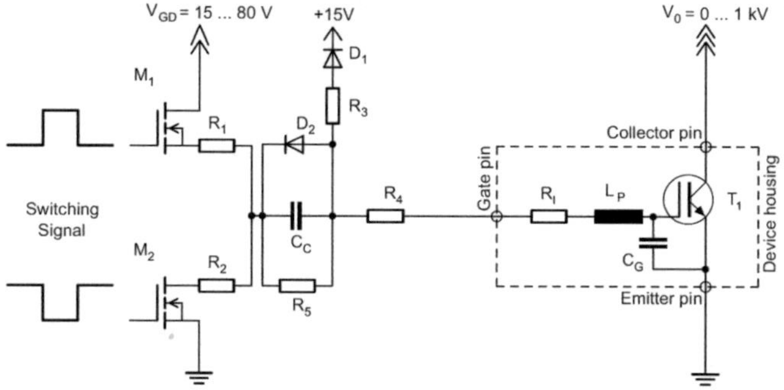

Figure 3.11: The gate boosting circuit under investigation.

The coupling capacitor C_C was adjusted to achieve 15 V in steady on-state following Eq. 3.4. For measurements with 15 V (i.e. "standard") gate drive voltage, the coupling capacitor was shorted. The equivalent gate capacitance C_C used in the formula is derived from experiments for each DUT as described in Sec. 3.3.3. The damping resistor R_D as introduced in the previous section is distributed over several discrete elements R_I, R_1, R_2, and R_4.

$$R_D = \begin{cases} R_I + R_1 + R_4; \text{ Charging of } C_G \\ R_I + R_2 + R_4; \text{ Discharging of } C_G \end{cases} \tag{3.10}$$

Although the remaining inductance in the connection of M_1 and M_2 is kept as low as possible, the very fast rise times of these switches can lead to a voltage overshoot across these devices during switching. The combination of R_1 and R_2 reduces oscillations across the drain-source capacitances of M_1 and M_2 and allows for a smaller safety margin when selecting the devices. As the circuit will be used for long pulses, charge loss has to be prevented. In steady-off state, any charge remaining on C_G will be shorted to ground via the high-value resistance R_5. In steady-on state, additional charge is supplied

via R_5. To prevent overcharging of the gate, excess charge is drained via R_3 and D_1. Since R_3 also has a high value, the voltage across R_4 and the gate pin can significantly exceed 15 V during transient turn-on. Parasitic turn-on as discussed in Sec. 4.2.1 is prevented by diode D_2, offering a low-resistance path to ground during turn-off.
Whereas there is no theoretical limitation of using higher gate drive voltages, there are several practical ones. The speed of the half-bridge switches will suffer with higher gate drive voltage, necessitating more expensive switches and/or a more complex drive circuitry. Additionally, Fig. 3.12 shows the dependence of the final gate voltage on the actual gate capacitance value of the respective device. For this plot, the circuit is adjusted to achieve 15 V gate voltage for a nominal gate capacitance of 10 nF. As can be seen, small part-to-part variations in the actual value of the gate capacitance have an increasing influence in the final gate voltage increases with increasing drive voltage (i.e. smaller coupling capacitor). Therefore, using gate drive voltages significantly above 80 V may result in time-consuming fine-tuning of the circuit for each respective device.

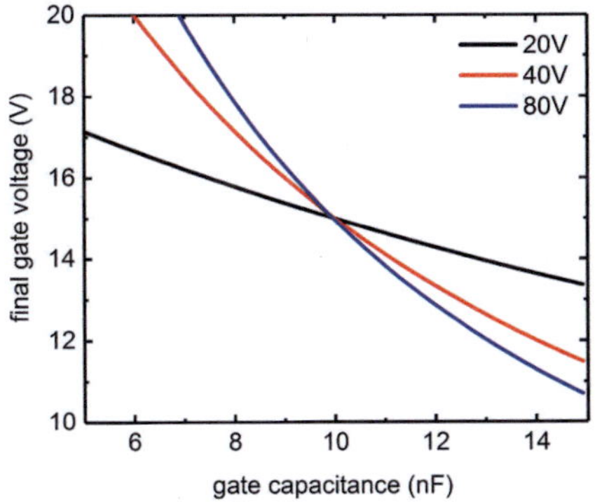

Figure 3.12: Final gate voltage in dependence of the actual value of the gate capacitance for different gate drive voltages. At high gate drive voltages, small part-to-part variations in the value of the gate capacitance become increasingly important.

Some manufacturers recommend an operation with bipolar gate voltage (i.e. ±15 V) [78]. In principle, the circuit can be easily adjusted to operate with bipolar voltage by shifting the potential of the M_2 source pin with respect to the T_1 emitter pin. However, this also increases the voltage across M_1 and M_2, resulting in the need for fast switches with a high blocking capability. As using a negative voltage mainly influences the turn-off delay and not the charge carrier recombination [64], the turn-off speed can only be influenced by this technique as far as a fast discharging of the gate capacitance is concerned. The potential difference across the coupling capacitor C_C during steady-on state (i.e. $V_{GD} = -15\,V$) propagates to the gate as transient negative drive voltage during turn-off. Therefore, the circuit translates the unipolar voltage of the driver into a transiently bipolar one and, hence, allows for a fast charging and discharging of the gate capacitance. The main benefit of a negative drive voltage during steady-off state being the increased robustness against parasitic turn-on [64], the aforementioned low-resistance connection to ground during turn-off is sufficient to prevent device damage in the intended application (see Sec. 4.2.1).

3.3.3 Determining the device parameters

For proper circuit design, knowledge about the parasitic device properties is crucial. Mainly the information about the gate capacitance, the parasitic inductance, and the internal gate resistance is not known a priori. While the datasheet usually contains a gate charge vs. gate-emitter voltage plot, this information is usually of limited use as it is given for one specific collector-emitter voltage only. Therefore, the straight forward method to determine the effective gate capacitance $C_{G,eff.}$ at nominal collector-emitter voltage in the design is to start with the value offered in the datasheet and measure the resulting gate-emitter voltage $V_{G,meas.}$ for a given coupling capacitor C_C and gate drive voltage V_{GD}. The effective gate capacitance of the device can then be calculated using Eq. 3.11.

$$C_{G,eff.} = C_C \cdot \left(\frac{V_{GD}}{V_{G,meas.}} - 1 \right) \tag{3.11}$$

Direct measurements of the parasitic inductance and internal resistance of the devices are challenging, as their values typically are very small. Comparing gate voltage measurements with simulations of the circuit (see Fig. 3.10), an iterative adjustment of the values yields sufficiently accurate results. As an example, an effective gate capacitance of 18 nF was determined for the

NGTB40N120 device operated at 1 kV collector-emitter voltage as compared to the nominal value of 10 nF. According to the simulation results, the parasitic inductance including the device leads and bond wires was around 8 nH, the internal gate resistance resulted to 0.8 Ω. If the gate drive circuit is to be operated in critically damped conditions, a minimum additional resistance of 0.6 Ω is necessary. However, for avoiding damage to the driving half-bridge, a minimum value of 1 Ω for all damping resistors was determined. Therefore, the gate drive circuit will operate in slightly overdamped conditions.

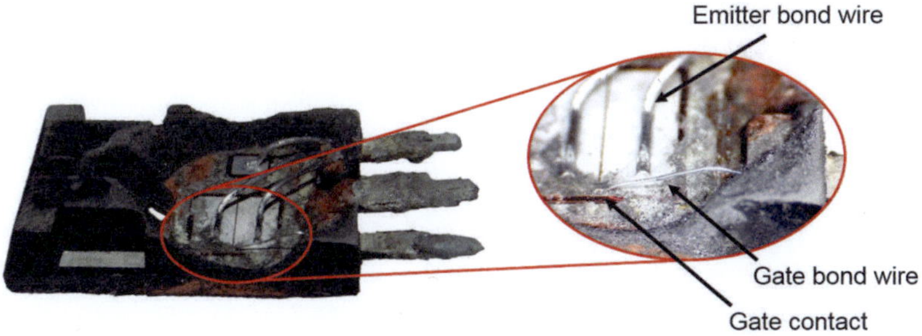

Figure 3.13: An IGBT in a commercial TO-247 housing, opened by an etching process to reveal the semiconductor dies of IGBT and co-packed diode.

The only way to prove the design process to be efficient is a measurement of the gate voltage directly at the die. To this end, an etching process with fuming nitric acid (above 90 % HNO_3, see [79]) was used to remove the protective epoxy housing around an IGBT die. The process can be significantly sped up by pre-heating the device in a petri dish on a hotplate to temperatures between 80 °C and 100 °C. By means of a pipette, drops of the acid are transferred to the heated device where the reaction starts immediately. Tilting the device, the acid can be used to wash the protective epoxy off the device die. After a sufficient amount of material is removed, the reaction is stopped by rincing the device with demineralized water.
A photograph of the resulting opened NGTB40N120 device can be seen in Fig. 3.13. As can be seen, the IGBT die consists of two separate chips, each contacted with a single emitter bond wire. The copper tab is used as collector contact at the backside of the die. The gate bond wire has a significantly smaller diameter than the emitter bond wires.

Using a low-inductive voltage probe, it is possible to measure the voltage along the gate bond wire (Fig. 3.14) when using 80 V gate drive voltage. Position 1 (black trace) is measured outside of the device as is possible without opening of the housing. As discussed before, the voltage signal shows a significant overshoot at this position. When tracking the signal towards the gate contact (red trace), the magnitude of the overshoot is reduced. Finally, measuring directly at the gate contact (Pos. 3, blue trace), the measured voltage shows a nearly rectangular pulse with no visible overshoot.

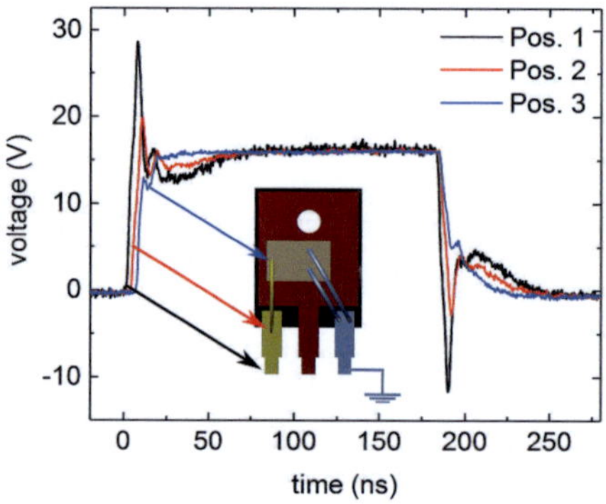

Figure 3.14: Gate-emitter voltage measurements on an opened NGTB40N120 device. The parasitic lead inductance causes a voltage spike as measured outside of the device, whereas a rectangular voltage pulse can be measured directly at the gate capacitance.

Thereby, the detrimental influence of the gate bond wires on the switching process is proven. One of the benefits of etching with fuming nitric acid is that the chip will remain functional. Due to the missing isolating coating, however, the blocking voltage is drastically reduced due to surface flash-overs.

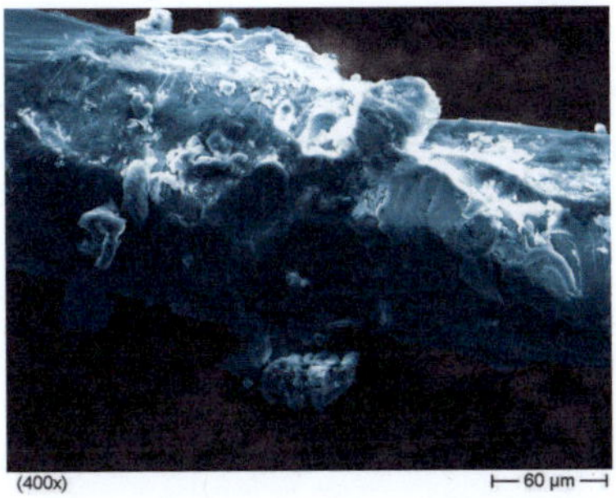

Figure 3.15: Colored SEM image of the bond wire showing a diameter of around 120 µm and moderate damage after the etching process.

A Scanning Electron Microscope (SEM) image of the gate bond wire (see Fig. 3.15) proofs that the aluminum wire is not severely damaged by the etching process, as a protective oxide layer is formed around it by the nitric acid. The diameter of the wire can be determined to 120 µm – the resulting intrinsic resistance of 24 mΩ considering the geometric dimensions of 1 cm and a resistivity of $2.8 \cdot 10^{-7}$ Ω m [41] can be neglected for the following calculations.

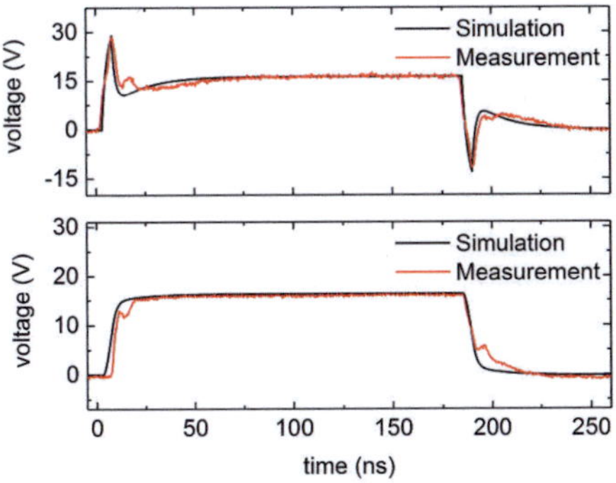

Figure 3.16: Comparison of gate voltage measurements and simulations. Using the parameters obtained from the iterative adaptation steps (compare Fig. 3.10), the externally measured signal and the signal measured across the internal gate capacitance are in good agreement with the simulations.

A comparison of the measurements shown in Fig. 3.14 (pos. 1 and 3) with the simulation results as presented in Fig. 3.10 is displayed in Fig. 3.16. The upper graph displays the measured and simulated traces outside of the device housing (i.e. pos. 1). As discussed before, the voltage overshoot due to the bond wire inductance is reproduced with good accuracy. Since the gate bond wire has no significant resistance, the obtained value of 0.8 Ω is part of the gate contact. Therefore, when measuring on the gate contact of the die, the internal gate resistance is part of the measurement. When simulating the voltage across the gate capacitance including the internal gate resistance and comparing it to the signal as measured at the IGBT gate contact (Fig. 3.16, lower graph), a reasonable agreement is achieved. The simulated voltage trace, hence, represents the voltage across the series connection of C_G and R_I as depicted in Fig. 3.11. By comparison of the upper and lower graph, it becomes clear that one reason of the turn-on and turn-off delay of voltage-controlled power semiconductors is caused by the time lag between voltage application at the device leads and the charging and discharging process of the gate capacitance. Nevertheless, the graph also proves the determination of the device parameters by iterative adjustment based on external measurements yields sufficiently accurate results for a rough determination of circuit parameters.

In order to validate the prior assumptions that the increase in gate capacitance caused by the Miller effect can be reduced to an increased effective gate capacitance, the IGBT was operated at 0 V collector-emitter voltage and 600 V collector-emitter voltage while the voltage across the internal gate capacitance was measured (compare Sec. 3.3.4). The measured voltage traces (Fig. 3.17) show no other relevant effects besides an increased effective capacitance causing a slightly lower final gate voltage and slower voltage rise. With the possibility of measuring directly across the gate contact (i.e. gate capacitance plus internal gate resistor), fine-tuning of the circuit is possible. As an example Fig. 3.18 shows a comparison between an adjustment for critically damped (black trace, $R_D = 4\,\Omega$) and underdamped (red trace, $R_D = 2\,\Omega$) circuit adjustment. The optimal conditions have to be determined for each type of device.

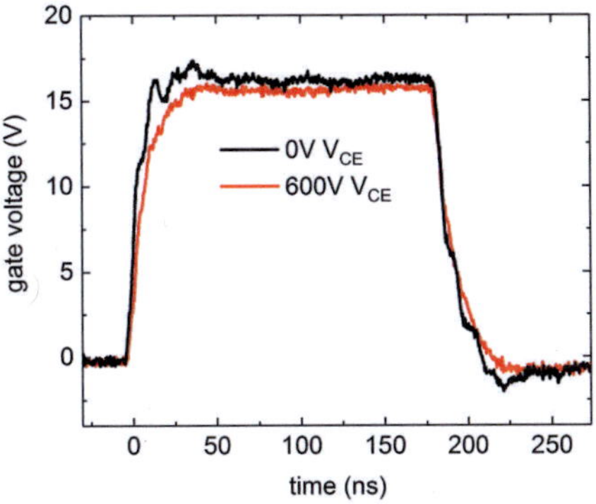

Figure 3.17: A comparison of the gate voltage as measured directly on the die for $V_{CE} = 0\,V$ (black trace) and $V_{CE} = 600\,V$ (red trace). The influence of the Miller-effect is visible, since the equivalent capacitance is increased for higher V_{CE}, decreasing the gate voltage.

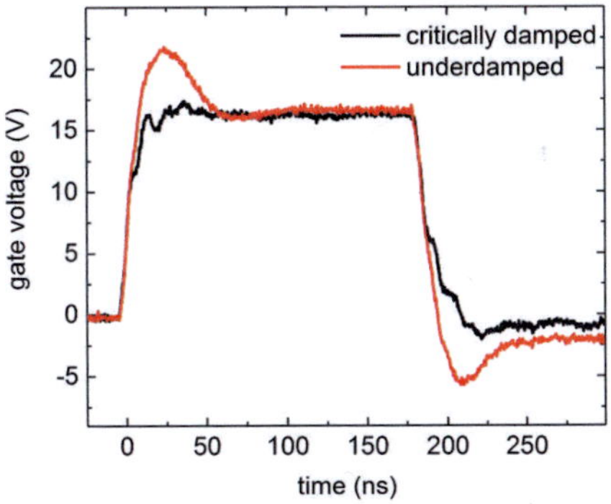

Figure 3.18: The gate voltage as measured directly on the gate capacitance when operating the circuit under critically damped and slightly underdamped conditions.

3.3.4 Increasing the switching speed

Pulse circuit qualification

To characterize the increase in switching speed for different devices when using the presented gate-boosting circuit, a low inductance pulse circuit (see Fig. 3.19) was used.
The DUT (T_1) was operated with varying gate drive voltage. The pulse circuit consists of a pulse capacitor C_1, a load resistance R_L, and a shunt resistor R_S. When closing the pulse switch, the voltage as measured across the switching element (P_1) drops while the current, measured as voltage drop (P_2) across the shunt resistor, increases. To avoid a voltage overshoot due to the remaining inductance in the load and shunt resistor, a free-wheeling diode D_1 is connected in parallel. A photograph can be seen in Fig. 3.20, identifying the circuit elements.

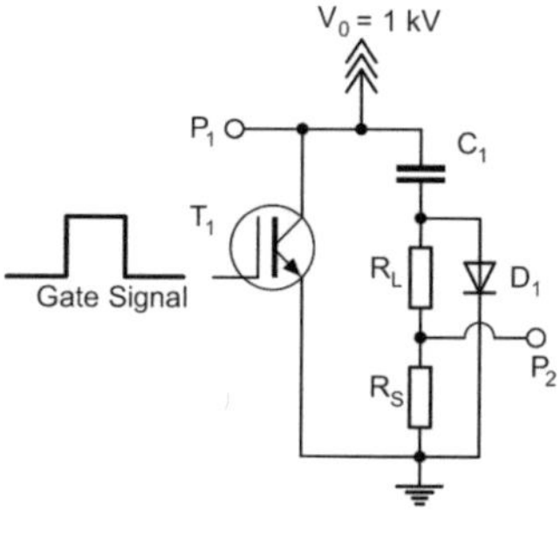

Figure 3.19: Schematic of the employed pulse circuit.

In order to achieve a fast rise time of the load current, a low-inductance arrangement is crucial. Therefore, the pulse capacitor C_1 consists of a combination of four low-inductance 1 µF (1) and two high value 100 µF (mounted on 2, not shown) capacitors. For a low-inductance load resistance R_L, several carbon-mass resistors (4) were connected in a parallel arrangement – the current was measured using a fast coaxial shunt resistor (3). The free-wheeling diode (5) and the DUT (6, on heat-sink) are mounted in the direction of current flow. All following data was recorded using a LeCroy WaveRunner 204MXi-A oscilloscope with 2 GHz (see Sec. 2.5).

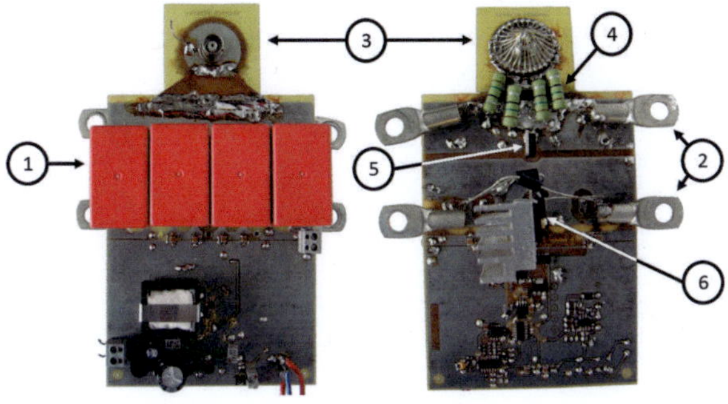

Figure 3.20: Photograph of the pulse circuit, showing the low-inductance capacitors (1), the mounting posts for two high value capacitors (2), the coaxial shunt resistor (3), the load resistors (4), the free-wheeling diode (5), and the DUT (6) mounted on a heat sink.

Using an LCR bridge (Hameg HM8118), the total circuit inductance was measured to be below 60 nH. In order to ensure hard switching conditions, the inductive component of the circuit must be sufficiently small. The current rise time constant τ as defined by Eq. 3.12 represents a lower limit for the current rise time in an LR circuit. For ohmic (i.e. hard switching) conditions, the current rise time of the switching element must be sufficiently higher than τ. The smallest load resistor used in the following measurements was 3.75 Ω, resulting in a time constant of 21 ns.

$$\tau = \frac{L}{R} \tag{3.12}$$

An example pulse of one NGTB40N120 with a collector-emitter voltage of 1 kV with 3.75 Ω load is shown in Fig. 3.21. With a current rise and fall time of approximately 50 ns, the absence of a phase lag between current and voltage signal is in agreement with the aforementioned calculations.

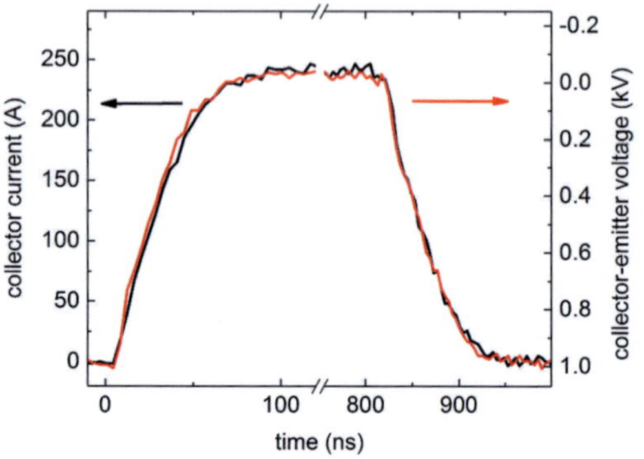

Figure 3.21: An example pulse with the employed pulse circuit shows the coincidence of voltage and current, validating the assumption of an ohmic load.

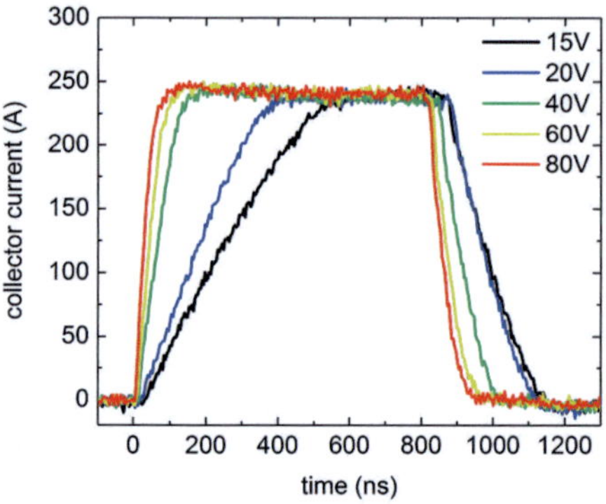

Figure 3.22: Collector current traces for different gate drive voltages. With increasing gate drive voltage, the switching speed of the NGTB40N120 device can be drastically increased.

Pulse measurements

In order to illustrate the changes in device performance under the influence of increasing gate-boosting voltage, measurements are presented for selected devices only. Using the circuit as described above, a Field-Stop Trench

IGBT (NGTB40N120IHLWG – ON Semiconductor [67]) was operated at 1 kV collector-emitter voltage connected to a resistive load of 3.75 Ω.
The device has a specified maximum pulsed current of $I_{C,max}$= 320 A and was operated at I_C= 256 A (80% I_{max}) collector current for a pulse length of 1 µs. As can be seen in Fig. 3.22, the increased charging speed of the gate capacitance associated with using higher gate drive voltages strongly increases the switching speed.
The associated collector current rise and fall times are shown in Fig. 3.23 and Fig. 3.25, respectively. For different maximum current amplitudes (i.e. 80 % $I_{C,max}$, 50 % $I_{C,max}$, and 25 % $I_{C,max}$), 100 measurements were conducted for each step in gate drive voltage. The corresponding error bars are in the order of several nanoseconds and, hence, are not visible in this scaling. As can be seen, the collector current rise time decreases strongly with increasing gate drive voltage. However, the extend of the speed-up seems to be depending on the current to be switched. For a current of 250 A, the turn-on speed increased from around 400 ns (standard gate drive voltage, i.e. 15 V) to 49 ns (using 80 V gate drive voltage).

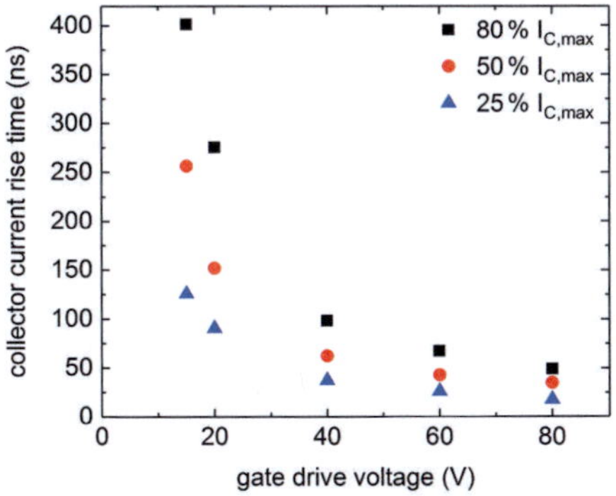

Figure 3.23: Reduction of current rise time for NGTB40N120 at different maximum current amplitudes as function of the gate drive voltage.

By normalizing the measurements of Fig. 3.23 to the standard (i.e. 15 V gate drive voltage) rise time for each current value, the relative decrease in rise time is obtained (Fig. 3.24).

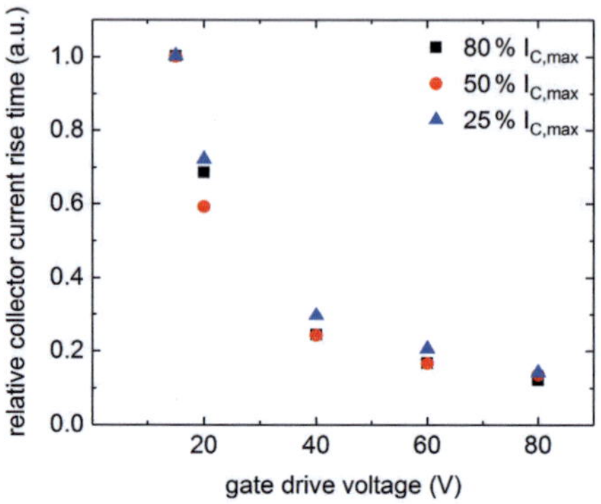

Figure 3.24: Reduction of the normalized current rise time for NGTB40N120 at different maximum current amplitudes as function of the gate drive voltage.

From this graph, it is obvious that the relative reduction in collector current rise time is nearly exponential and, in good approximation, independent of the pulse current. This type of exponentially-like decrease is in agreement with the predictions derived from the analytical analysis of the circuit according to Sec. 3.3.1 as a result from an increased gate charging speed. For all current values, the increase in switching speed is nearly eight-fold. Additionally, the relative deviation in current rise time between high and low pulse current is kept nearly constant, allowing for an equal current sharing during turn-on as discussed in Sec. 4.1.1.

Considering the trailing edge of the pulse (i.e. the current fall times), the picture changes slightly. Fig. 3.25 shows collector current fall times in dependence of the gate drive voltage. For high pulse currents, the decay in current fall time with increasing gate drive voltage is visible if not as pronounced as for the current rise times.

Interestingly, the current rise time for very low current (blue symbols) is influenced less strongly by the gate-boosting circuit. As figure of merit for the increase in switching speed, the current rise rate is often employed.

In Fig. 3.26, the current rise and fall rate of the above measurements (definition see Sec. 2.5) is plotted. Whereas the current fall rate at low currents (red triangle) cannot be influenced as mentioned above, especially the current rise rates (black symbols) show a linear dependence on the gate drive voltage.

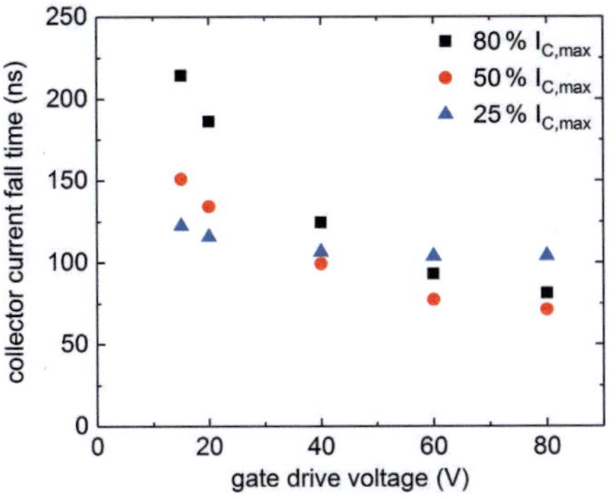

Figure 3.25: Reduction of current fall time for NGTB40N120 at different maximum current amplitudes as function of the gate drive voltage.

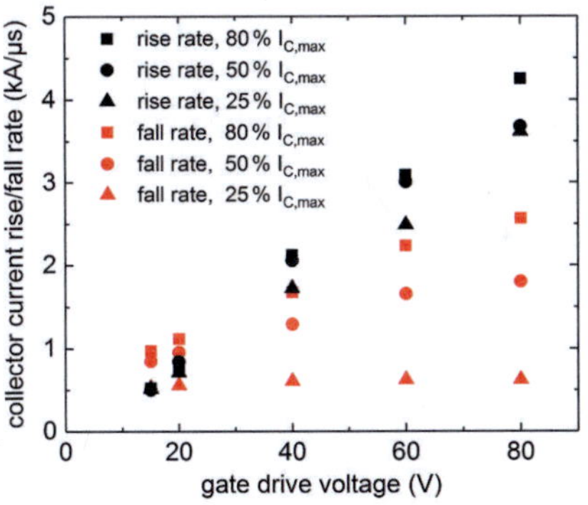

Figure 3.26: Collector current rise and fall rates at different maximum current amplitudes in dependence of the gate drive voltage. The increase in current rise rate with higher gate drive voltages is still linear for the maximum tested voltage, suggesting even further improvements with high gate drive voltages. The impact on the current fall rates is less pronounced.

The linear dependence of this figure of merit is not surprising since the reciprocal function of an exponential decay is linear for small values. For low gate drive voltages, the current fall rate is higher than the current rise rate. As

the slope of increase is higher for the current rise rates, the device turns on faster as it turns off for gate drive voltages above 40 V. Since no saturation effects are visible at high gate drive voltages, it is reasonable to assume that the investigated effect can be used beyond the maximum tested 80 V gate drive voltage for achieving much higher collector current rise rates. With the current rise time being fast enough for the intended application, the increased costs associated with using even higher drive voltages were avoided by limiting the experiments to 80 V.
With the now increased switching speed for turn-on and turn-off, also the maximum operating frequency can be drastically increased. Implementing the gate-boosting circuit, Fig. 3.27 shows the operation of the investigated NGTB40N120 device in 3 MHz burst mode operation at approximately 250 A collector current.

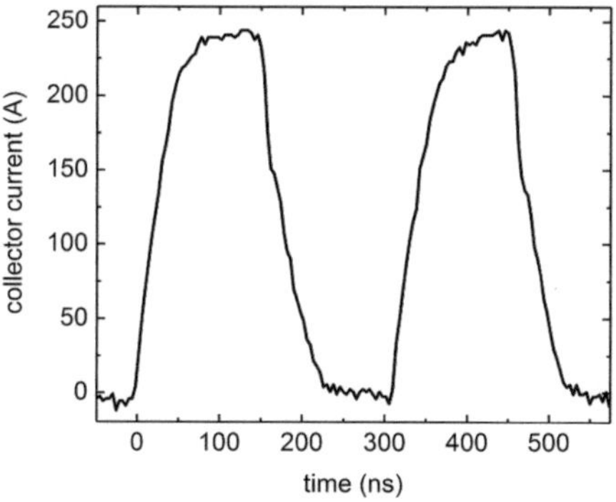

Figure 3.27: By using the gate-boosting circuit, a standard IGBT can be operated at approximately 250 A with 3 MHz bursts.

In addition to the presented measurements on the NGTB40N120 IGBT device, several other devices have been qualified in the course of this work.
An interesting example is the SiC MOSFET C2M0080120D manufactured by Wolfspeed. The device is able to carry a maximum of 80 A pulsed current at a maximum blocking voltage of 1200 V. Therefore, it could be used in the test-bed as described earlier.

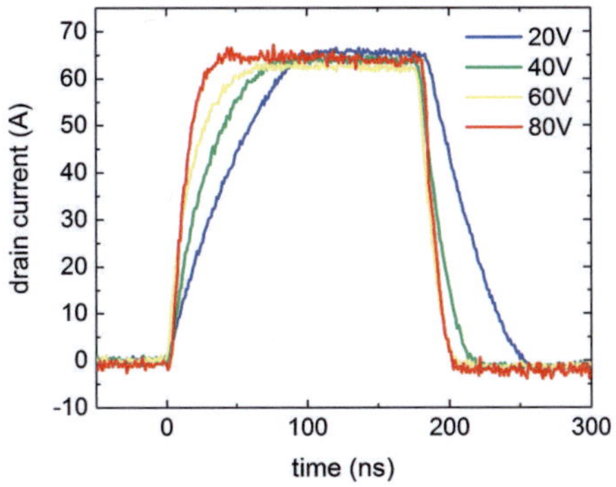

Figure 3.28: Pulse current measurements for a SiC MOSFET at different gate drive voltages, showing the applicability of the technique also for other device architectures.

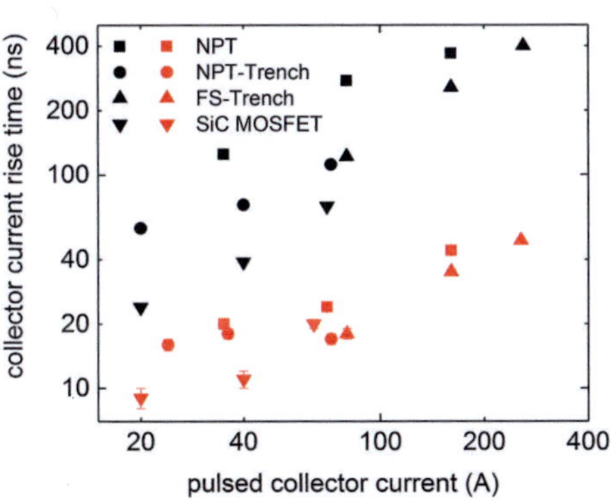

Figure 3.29: Comparison of the increase in switching speed for different devices (see table 3.1) and architectures (black symbols: standard gate drive conditions, red symbols: 80 V gate drive voltage).

When operating it under the same conditions as the IGBT device (i.e. 80 % maximum pulsed current, 65 A), graph 3.28 could be obtained for increasing gate drive voltage. Here, 20 V are the recommended operating conditions used with the coupling capacitor shorted. As for the IGBT device, the switching speed drastically increases with increasing gate drive voltage. Thereby, these

measurements prove the applicability of the gate-boosting circuit for the most commonly used voltage-controlled power semiconductors, MOSFETs and IGBTs. In order to compare the effect on different devices, data of all measured devices is combined in Fig. 3.29.
Black symbols display values measured in standard gate drive conditions, red symbols display values at 80 V gate drive voltage. For each device, measurements were taken at 25 %, 50 %, and 80 % maximum pulsed current. Each point represents the mean value of 100 measurements, the error bars are in the order of several nanoseconds and therefore not visible except for very fast switching speeds. All devices are summarized in table 3.1.

Table 3.1: Different devices investigated in this work. [80], [81], [67], [61]

Device architecture	Device part number	Max. pulsed current	$t_{r,Std}/t_{r,80V}$
NPT IGBT	HGTG18N120	160 A	8.4
NPT Trench IGBT	FGA25N120ANTD	90 A	6.6
FS-Trench IGBT	NGTB40N120IHLWG	320 A	8.2
SiC MOSFET	C2M0080120D	80 A	3.6

As expected, the collector current rise time increases for all devices with increasing pulse current. Only for the NPT-Trench device, this behavior is changed by using the gate-boosting circuit. From the graph it becomes clear, that the change in switching speed from one device to the next does neither follow exactly the same shape nor has the same magnitude. In the last column of table 3.1 the relative increase in switching speed (i.e. rise time under standard drive conditions over rise time using 80 V gate drive voltage – both at maximum tested current) is calculated. The values differ quite significantly from one device to the next. Therefore, it is necessary to investigate each device with respect to the intended application prior to device selection.
As discussed in Sec. 3.3.1, one way to explain the working principle of the gate-boosting circuit is the high current rise rate through the parasitic gate inductance that can be achieved using higher gate drive voltages. As a result, the drive circuitry becomes increasingly robust against parasitic inductance introduced by imperfect gate circuit layout. To demonstrate the effect, a device was operated using 80 V gate drive voltage with the standard gate circuit (black trace, Fig. 3.30). For another measurement, the parasitic gate inductance was

increased by approximately 10 nH (red trace, Fig. 3.30). In comparison, the additional inductance of 10 nH was moved to the emitter pin (compare L_E, Fig. 3.4b) of the switching device (green trace, Fig. 3.30). From the traces it is evident, that the gate-boosting circuit can compensate parasitic gate inductance to a certain extend. However, already a small parasitic inductance shared between gate and load circuit deteriorates the pulse shape. Therefore, shared current paths of both circuits have to be avoided for maximum efficiency of the circuit (compare Sec. 3.2.1).

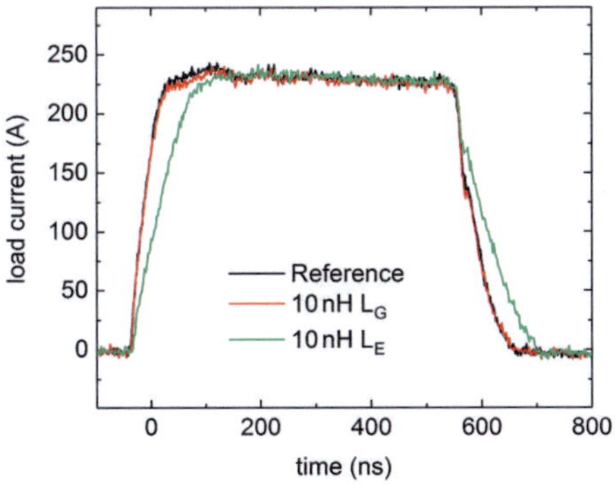

Figure 3.30: Load current traces for the reference circuit, and the effect of additional 10 nH inductance in the gate (red trace) and emitter (green) circuit.

3.3.5 Reduction of switching losses

When using power semiconductors at high switching frequencies, the switching losses can easily deliver a more important contribution to the overall losses than the conduction losses [64]. The energy loss during each switching event (from t_0 to t_{end}), being defined according to Eq. 3.1, can be kept small by introducing a phase shift between current and voltage and then switch when either current (zero-current switching, ZCS) or voltage (zero-voltage switching, ZVS) across the switching element is zero (soft switching). For hard switching conditions, necessary to achieve a fast current rise rate, the only way to reduce switching losses for constant switching power is to decrease the transition time of the switch from on- to off-state.

For the measurements in graph 3.31, the pulse current and voltage traces as presented in Fig. 3.22 were multiplied. As can be seen, the losses occurring during turn-on greatly exceed the losses during turn-off using the standard gate drive circuit (black trace). When increasing the gate drive voltage, the overall switching losses are strongly reduced, with the reduction in turn-on losses being far more pronounced than the reduction in turn-off losses as the turn-off speed can not be altered as significantly (see Sec. 3.3.4).

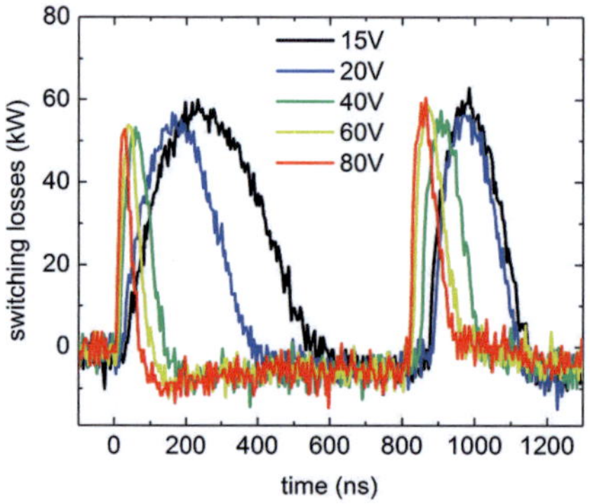

Figure 3.31: Decrease of switching losses for a NGTB40N120 device with increasing gate drive voltage.

The energy loss E_{Loss} during turn-on and turn-off normalized to the energy loss using the standard gate drive circuit follows the exponential decay observed for the decrease in current rise time as presented earlier (see Fig. 3.32).
As a result of the energy being a product of current and voltage (both of which are sped up), the decrease in switching losses is even more drastic. By using 80 V gate drive voltage, the turn-on losses can be reduced by a factor of 10, whereas the turn-off losses can be reduced by a factor of 2.5.
As mentioned before, in certain applications a reduction of switching losses is achieved by resonant switching and, therefore, the drive losses will dominate the switching losses.

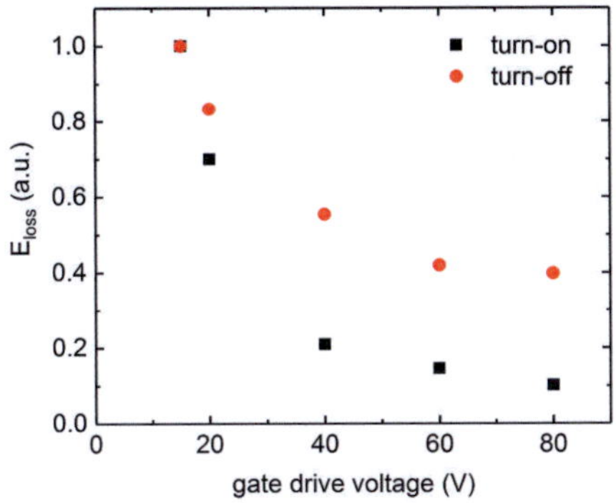

Figure 3.32: Energy loss, normalized to the energy loss using the standard gate drive circuit, during turn-on and turn-off when employing the gate-boosting circuit with 80 V gate drive voltage.

The energy E_G to charge the gate capacitance of a voltage-controlled power semiconductor is proportional to its gate capacitance C_G and the square of the gate voltage V_G.

$$E_G = \frac{1}{2} \cdot C_G \cdot V_G^2 \tag{3.13}$$

When using the gate-boosting circuit, the equivalent capacitance C taking the place of the gate capacitance in Eq. 3.13 is reduced whereas the gate drive voltage V_{GD} is increased. By combining Eq. 3.6 and Eq. 3.4, the equivalent capacitance can be re-written.

$$C = C_G \cdot \frac{V_G}{V_{GD}} \tag{3.14}$$

Thereby, Eq. 3.13 yields the dependence of the required energy for one switching event in dependence of the chosen gate drive voltage.

$$E_G(V_{GD}) = \frac{1}{2} \cdot C \cdot V_{GD}^2 = \frac{1}{2} \cdot C_G \cdot V_G \cdot V_{GD} \tag{3.15}$$

Therefore, the required drive energy (and drive power) grows linearly with increasing the gate drive voltage. For a complete loss analysis, also losses accruing in the driving half-bridge and the efficiency of the DC-DC converter generating the gate drive voltage have to be considered. Since those parameters are depending on the actual circuit design, these calculations have to be done specifically for each design.

3.3.6 Long term stability

The previous discussion concerning the operation of the device within the datasheet specifications was limited to not exceeding the maximum specified gate voltage.

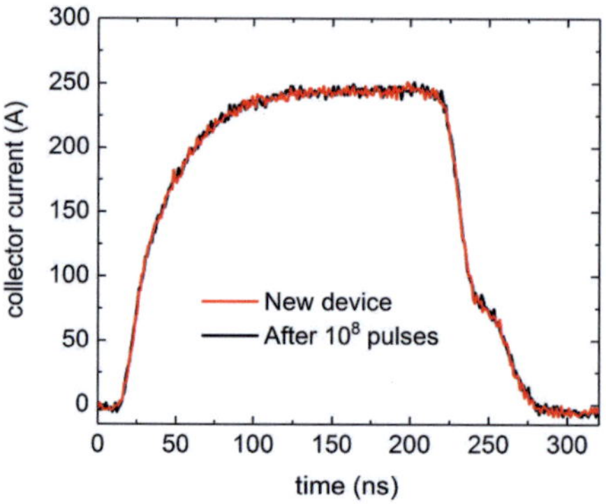

Figure 3.33: Pulse shape of a new device and a device after 10^8 pulses.

However, additional effects may cause a slow device degradation when using the gate-boosting circuit. Among these might be the increased thermal stress on the gate bond wire, potentially causing a bond wire lift-off. Strongly increasing the collector voltage rise rate, the increased switching speed may even cause internal device latch-up [62]. In order to evaluate the long-term effects of the gate-boosting circuit on a device, the setup was upgraded to operate at a repetition rate of 3 kHz. At maximum gate drive voltage (i.e. 80 V), the NGTB40N120 device was operated at maximum collector-emitter voltage (i.e. 1 kV) and maximum pulsed current (i.e. 250 A) at a pulse length of 250 ns. In order to cope with the increased power dissipation at high repetition rates, the IGBT was mounted on a water-cooled heat sink with a thermal resistance of 0.18 K/W (case-to-water). Together with the thermal resistance of the package of 0.48 K/W (junction-to-case) [67], the total thermal resistance equals 0.66 K/W (junction-to-water). In thermal equilibrium, the water temperature at the outlet was measured to be 23 °C, whereas at the inlet it was chilled to 18 °C. The resulting power dissipation in the switching element can be calculated to approximately 70 W. The junction temperature of the die during the measurement was, hence, around 70 °C and therefore much above the maximum expected temperature in the generator application operating in single pulse operation. Under these conditions, the device was operated for 10 h, being equal to more than 10^8 pulses. A comparison of the pulse shape of a pristine device and a device after the test is shown in Fig. 3.33 – both traces are identical within the resolution of the measurement. Whereas for standard semiconductor applications the tested pulse number seems not impressive, the device was tested far beyond the requirements of the intended application. The evaluation of the technology at industry level will therefore remain object of further studies.

4 Stage design process

The following section describes the design process of one stage as the basic element of the modular Marx generator. Besides the selection and arrangement of the pulse circuit components, also related protection circuits in different modes of operation are described. Concerning the control logic on each stage, this section covers the hardware side of the optical signal distribution, employed logic devices, and the challenge of shielding them against electo-magnetic interference. The software and the interaction of many stages is presented in Sec. 5. All design parameters are validated in a low-inductive test-bed demonstrating the maximum output parameters of 1 kV and 600 A in arbitrary switching sequences. Major results of this work presented in the following chapter have already been published as [82].

4.1 Pulse circuit components

One stage of the generator-under-development consists of a pulse switch T, a stage capacitor C and a free-wheeling diode D (see Fig. 4.1). In all circuit diagrams, "next stage" refers to the stage closer to the generator output and "previous stage" symbols a stage closer to the grounding point of the generator.

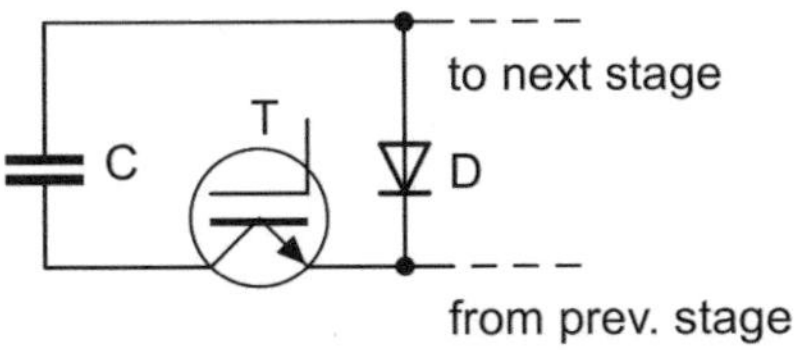

Figure 4.1: The pulse circuit components of one stage include the pulse switch T, the free-wheeling diode D and the pulse capacitor C.

4.1.1 Fast, cost-efficient switches

From the results of the previous section, the most economical solution was found to be the employment of high-power IGBTs using a gate-boosting circuit as switching elements on each stage. As mentioned in Sec. 3.1.2, the use of IGBT modules is not economical. Also unsymmetrical power lead layout within the module may lead to unequal current sharing between the IGBT chips and, consecutively, to a destruction of the module under high pulsed currents [75]. The IGBT type NGTB40N120 (ONSemiconductor) was selected due to its low price and good switching performance with respect to maximum current and achievable current rise time when using a gate-boosting voltage of 80 V. To achieve the maximum required pulsed current of 600 A, the straight-forward solution is to parallel many individual devices. Beside a symmetrical layout for each switch, also intrinsic parameters have to be considered. All newer devices based on NPT or FS-trench technology feature a positive temperature coefficient to prevent thermal run-away of single devices [64]. Hence, thermal coupling is not as important as it was in the earlier days of IGBTs. In the course of the experiments in the previous section (compare Fig. 3.23) the current rise time was proven to increase with higher currents. This benefits equal current sharing between many devices during transient turn-on, as a device starting to turn on with a small delay (and therefore carries a smaller current) turns on faster than the devices carrying a high current from the beginning. The strong asymmetry of turn-on and turn-off times and their dependence on the current is no issue for this application, as the only strong requirement considering switching speeds is the turn-on time. Aiming for a final current rise time of the generator in the order of 4 kA/µs limited only by the circuit inductance, the switches only need to be sufficiently faster as not to influence the cathode charging process (compare Sec. 2.3.1). The selected NGTB40N120 device is able to carry a maximum pulsed current of 320 A at a maximum DC current of 80 A [67]. According to the Safe-Operating-Area (SOA) in the datasheet, the maximum pulsed current can be applied for a duration of 1 ms. Allowing for a sufficient safety margin in case of an overcurrent condition, six devices were used in parallel to make up the stage's pulse switch. Thereby, each device carries a current of 100 A at maximum rated current, while the generator can withstand a maximum current of 1.9 kA for a duration of 1 ms. Using even more switches at lower current amplitudes also increases the tail-current, thereby skewing the falling edge of the pulse.

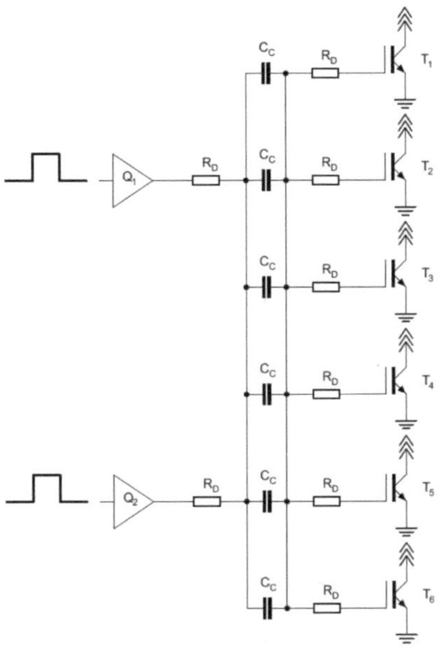

Figure 4.2: Simplified schematic of the employed capacitively coupled gate-boosting circuit as described in Sec. 3.3 for parallel operation of six IGBTs (T_1 to T_6) using two driver units Q_1 and Q_2. For equal stray inductances, each device has its own coupling capacitor C_C and damping resistor R_D.

To foster a synchronous turn-on of all IGBTs, the gate drive circuit was set up according to Fig. 4.2. The number of individual driver units Q is chosen according to the maximum current one unit can deliver. For the given values[1] of damping resistors R_D, gate drive voltage V_{GD}, gate and coupling C_C capacitance, the gate current of one device is around 13 A. The gate driver being able to deliver a peak current of 40 A leads to the use of two individual gate driver units. By interconnecting both sides of the coupling capacitors the influence of delayed switching of one driver can be mitigated. The drivers are located as close as possible to the switches, minimizing stray inductance. Fig. 4.3 shows a photo of the gate drive arrangement for six parallel IGBTs. As the employed TO-247 housing features no Kelvin-emitter, the separation

[1] $R_D = 1\,\Omega$, $V_{GD} = 80\,V$, $C_G = 18\,nF$, $C_C = 4.7\,nF$

of load and gate drive current as discussed in Sec. 3.3.4 can only be achieved outside of the device housing. Therefore, the gate drive circuit is located on one side of the devices, while the load is connected via the other side. The collector trace is a broad-side trace on the top layer, whereas the emitter trace is a broad-side trace on the bottom layer. By bending the IGBTs by 90° flat on the board, magnetic coupling between the device leads and the current traces will further reduce the inductance of the setup, while additionally reducing the mechanical dimensions. An illustration is shown in the following section (Fig. 4.5).

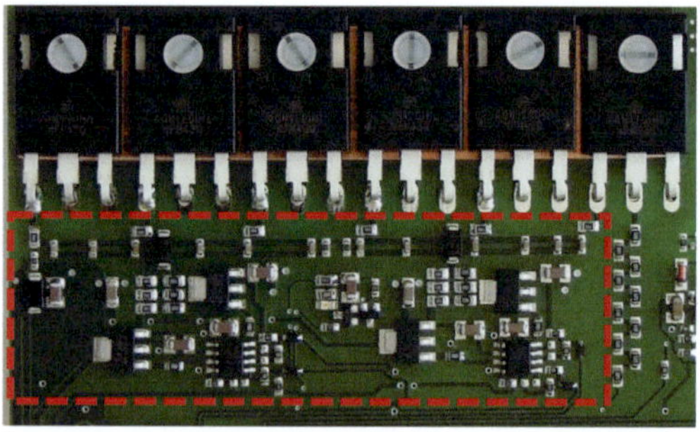

Figure 4.3: Photo of the two gate drive units on the stage for driving six IGBTs in parallel.

4.1.2 Pulse capacitors and capacitor connection

Besides the switching elements, the energy storage on each stage needs to be fast enough as not to deteriorate the output waveform. low-inductance pulse capacitors are available as metalized film capacitors[2] or ceramic capacitors[3], but their price per stored energy is very high. As outlined in Sec. 2.3.3, each stage requires a considerable capacitance of approximately 200 μF. Therefore, the use of the aforementioned capacitor types is not economical for storing all the required energy. The presented design in this work instead uses two metalized film capacitors intended for DC-link applications[4]. To compensate

[2] See e.g. MKP10, WIMA [83], 10€/J
[3] See e.g. C0G SMD pulse capacitors, KEMET [84], 640€/J
[4] 944U101K122AC, CDE [85], 1€/J

their high internal inductance of 40 nH, four additional pulse capacitors[5] with 1 µF each with an internal inductance of approximately 1 nH [83] are placed in close vicinity to the switching elements. Thereby, also the spacial dimensions of the energy storage could be drastically reduced as compared to an energy storage of purely fast pulse capacitors. A photograph of the arrangement can be seen in Fig. 4.4.

Figure 4.4: Photo of the presented Marx-generator stage. The switching elements and the capacitors are connected via broadside traces across the whole PCB width.

The inductances between the two types of capacitors and the capacitors and the switching elements are kept low by using broad-side traces over the whole width of the PCB. Fig. 4.5 illustrates the capacitor and switch arrangement. The high-value, high-inductance 100 µF capacitors (gray) are mounted on the right-hand side of the PCB and connect to its top and bottom layer by means of high surface area copper-beryllium contact springs. Another benefit of this mounting technique is the mechanical decoupling between the pulse capacitors and the PCB. The high-capacitance capacitors are connected to the low-inductance pulse capacitors (red) via a trace covering the whole top and bottom layer. The red capacitors are spaced evenly across the width of the PCB and are mounted through the PCB to offer a low-inductance connection. As can be seen, the connections from and to following stages are in close vicinity to one another, only separated by the free-wheeling diode. The pulse

[5] MKP1T041007H

current, entering and exiting from below therefore flows anti-parallel on the broad top and bottom layer of the board, allowing for an efficient magnetic flux compensation and a reduced inductance.

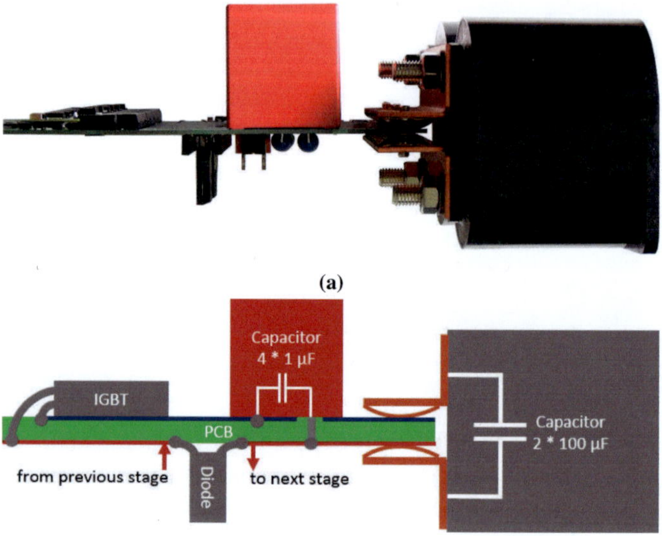

(a)

(b)

Figure 4.5: Photo (a) and illustration (b) of the component arrangement on one stage. The high-value 100 µF capacitors (gray) are mounted on the right-hand side of the PCB and are connected to the switching elements by means of broad-side traces over the whole width of the PCB. The low-inductance pulse capacitors (red) are mounted through the PCB to reduce the stray inductance. The connections from and to the adjacent stages are very close together.

The additional stray inductance per meter introduced by this arrangement can be estimated using Eq. 4.1 [15].

$$L' = \mu \frac{d}{w} \tag{4.1}$$

Using the dimensions of the board[6], the stray inductance can be calculated to 2.4 nH. As a result, the equivalent circuit of one stage is modified according to Fig. 4.6. The stage capacitor C now consists of a high-value capacitance C_1 with a high-parasitic inductance L_1 (as parallel connection of two 100 µF capacitors) and low-value capacitor C_2 with a low parasitic inductance L_2 (as parallel connection of four 1 µF capacitors).

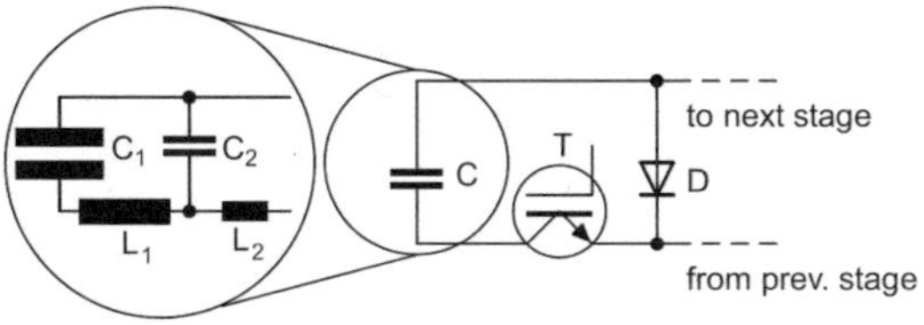

Figure 4.6: Equivalent circuit of one stage. The stage capacitor C is comprised of several individual, ideal capacitors and their respective series inductance.

The small value capacitors serve two purposes. On the one hand, they supply the charge necessary to create a fast rising pulse before the current commutates to the high-value capacitors. On the other hand, they serve as snubber protecting the switching element in case of current interruption as they can absorb the energy stored in L_1 during a nominal pulse (see Sec. 4.2.2). However, the resonant circuit formed by L_1 and C_2 causes charge oscillations between the capacitors. To investigate this phenomenon, the circuit was simulated in SPICE according to the equivalent circuit presented in Fig. 4.6[7]. Damping was incorporated to account for ohmic, dielectric, magnetic, and radiated losses. A comparison between the simulated voltage across the switching element and a corresponding measurement is shown in Fig. 4.7. For this measurement, a stage was operated at maximum pulse parameters of 1 kV charging voltage and 600 A pulse current (measurement without additional snubber circuit, for details see Sec. 4.2.2). After turn-off, the voltage overshoot is below 1060 V, demonstrating the efficient buffering of the energy stored in L_1 in C_2. The visible charge oscillations after turn-off correspond well to the resonant frequency of L_1 and C_2 of around 470 kHz. The strong damping prevents a significant

[6] width w = 10 cm, distance d = 1.5 mm; length l = 13 cm
[7] C_1 = 200 µF, L_1 = 30 nH, C_2 = 4 µH, L_2 = 2 nH

voltage overshoot even at an excitation at resonant frequency as one stage is able to switch with frequencies above 500 kHz (see Sec. 4.4.3). The additional current stress on the capacitors has to be considered in applications running at high repetition rates, but can be neglected for single pulse experiments.

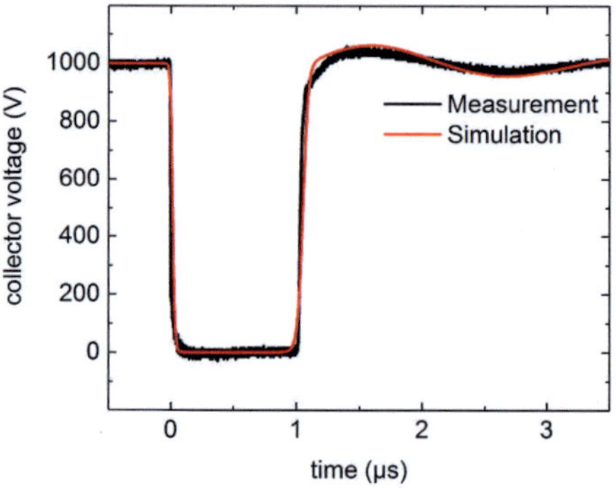

Figure 4.7: Comparison between measurement (black trace) and simulation (red trace) of the collector-emitter voltage across the switching element in the course of a short pulse. After turn-off, charge oscillations between the stage capacitors are visible. There is no significant voltage overshoot during turn-off.

4.1.3 Arrangement and selection of free-wheeling diodes

The free-wheeling diode on each stage is crucial for arbitrary output waveforms as it takes over the pulse current as soon as the pulse switch opens. For proper functioning, its position with respect to the switching element and the pulse capacitor plays an important role. As can be seen in Fig. 4.8, the closer the diode D is moved towards the switching element T and pulse capacitor C, the smaller the parasitic inductances $L_{P'}$ and $L_{P'''}$ become. When the switch opens, all inductances are energized. Only the energy stored in $L_{P''}$ and $L_{P''''}$ can be diverted to the next stage whereas the energy of $L_{P'}$ and $L_{P'''}$ has to be dissipated on the stage. Therefore, the diode has to be positioned as close as possible to both contacts connecting the stages.

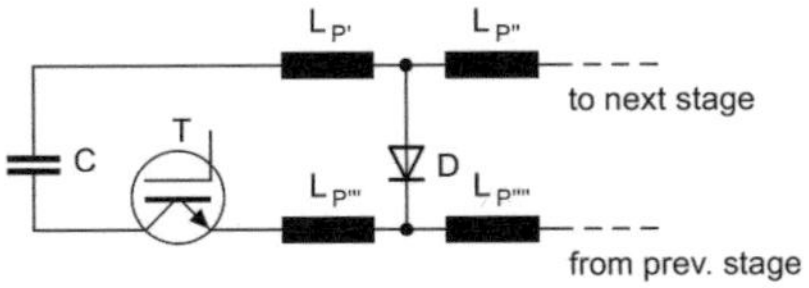

Figure 4.8: The position of the free-wheeling diode D with respect to the pulse capacitor C and the pulse switch T determines the distribution of the stray capacitances L_P. Only the energy in the stray capacitances $L_{P''}$ and $L_{P''''}$ can be diverted by the free-wheeling diode.

The chosen arrangement for the diodes can be seen in Fig. 4.5. Surface-mounting the diodes is unconventional, but offers superior performance in this application due to the absence of current trace interruptions that would otherwise occur when using a through-the-hole mounting technique. The diodes employed as free-wheeling diodes must fulfill three requirements: first, they have to be capable to block the stage charging voltage of 1 kV. Secondly, they have to carry a pulse current of up to 1.9 kA, which is the maximum capability of the switching elements. Most importantly, the use of high-speed switching elements also necessitates a very fast blocking behavior. Employing bipolar p-n junction diodes has the disadvantage of considerable reverse recovery times related to the recombination of charge carriers within the junction [63]. P-n diodes are considered "hyperfast" with recovery times in the order of 84 ns [86]. Taking the achievable current rise rates of 4 kA/µs for single devices into account, the maximum current rise rate of six devices in parallel amounts to 24 kA/µs given a low-inductance arrangement. Within the 84 ns reverse recovery time of a p-n junction free-wheeling diode, a current rise to 2 kA can be expected. Additionally, some p-n junction devices have a significant forward recovery time, rendering them unsuitable for this application [87]. High-band gap Schottky-barrier devices, such as SiC-Diodes, have the advantage of zero reverse recovery current due to the unipolar current transport [63]. The stress on the switching elements is thereby limited to the charging current of the diode's capacitance. Their increased reverse current as compared to Si p-n devices [63] plays a minor role in this application. The diodes chosen for this application are of type C4D10120D from Wolfspeed [88]. With two individual diodes (i.e. "legs") incorporated in one TO-247 housing, they offer a compact layout and high pulse handling capability. By correcting the values given in the datasheet for square pulses, each leg of one device can handle 100 A for a duration of 100 µs. Combining three devices (i.e. six diodes), the pulse specifications can be met. With a short term overload capability of 400 A for 10 µs, the arrangement exceeds the total current rating of the IGBTs: 2.4 kA for all diodes versus 1.9 kA for the IGBTs.

4.2 Protection circuits

Failure of power semiconductor devices is either related to a permanent alteration of their crystal structure or damage to their bonding. Both types are caused by subjecting parts of the device to excessive heat. Therefore, both current and voltage must be present for a given amount of time [89]. If the respective process is fast compared to the heat transfer within the device – as is the case during a breakdown for example – the limit is the adiabatic energy the device can absorb. For slower processes, heat transfer has to be considered and the maximum permissible current voltage product can be derived from the SOA diagram in the datasheet.

4.2.1 Parasitic turn-on

Following the discussions on the Miller-effect in Sec. 3.2.1, a parasitic turn-on can happen when the collector-emitter voltage across the switching device rises too fast. The collector-gate capacitance $\mathrm{C_{GC}}$ and the gate-emitter capacitance $\mathrm{C_G}$ act as a capacitive divider (see circuit diagram Fig. 3.4). For slower transients, the device can turn on when the voltage drop over the gate resistor required to charge $\mathrm{C_{GC}}$ in combination with the induced voltage over the parasitic circuit inductance $\mathrm{L_P}$ by the $\mathrm{C_{GC}}$ charging current $\mathrm{I_{GC}}$ is higher than the gate threshold-voltage $\mathrm{V_{G,Thres}}$:

$$\mathrm{V_{G,Thres}} < \mathrm{R_G} \cdot \mathrm{I_{GC}} + \mathrm{L_P} \cdot \frac{\mathrm{dI_{GC}}}{\mathrm{dt}} \tag{4.2}$$

Robustness against parasitic turn-on is especially important in a half-bridge configuration [90]. When operating the IGBT as switch in the Marx configuration investigated in this work, the voltage across the IGBT is in off-state almost equal to the voltage across the stage capacitor. The free-wheeling diode will clamp the voltage difference to several volt. A fast voltage change across the IGBT would, hence, require a very high charging current into the stage capacitor. Additionally, the implemented gate-boosting circuit offers a low-impedance connection from gate to ground in off-state, reducing the first contribution in Eq. 4.2. The second contribution is kept low by a short connection between gate contact and gate driver, reducing L_P.

4.2.2 Overvoltage protection

In the investigated topology, overvoltage conditions for the switching elements occur only as inductively driven voltage spikes when large currents are being cut-off. Due to the fast switching speed of the employed power semiconductors, the energy stored in the parasitic inductances cannot be dissipated during switching and cause a voltage peak across the device. The inductive overvoltage V_O can be calculated from the load current change rate dI_L/dt and the parasitic circuit inductance L_L.

$$V_O = L_L \cdot \frac{dI_L}{dt} \tag{4.3}$$

By exceeding the voltage blocking capability of the device, avalanche breakdown occurs. Whereas there are devices rated for avalanche operation available, the maximum energy they can absorb prior to device failure is limited [30]. The energy E_P stored in the parasitic inductance L_L at load current I_L equals:

$$E_P = \frac{1}{2} \cdot L_P \cdot I_L^2 \tag{4.4}$$

Only few IGBTs are rated for avalanche operation, since their switching speed (without using a gate-boosting circuit) is too slow for important inductive voltage spikes to occur [91]. In contrast to the current conduction in on-state, the current distribution during avalanche breakdown is not uniform but is concentrated on the perimeter of the die [91]. As the physical device housing often incorporates a free-wheeling diode on a smaller die, the diode absorbs all the energy during avalanche and therefore fails first [91]. In experiments, excessive overvoltage gradually degrading the device could be measured by an increased leakage current.

As measured in Sec. 4.1.2, the combination of different capacitor types allows to cope with the energy stored in the capacitor inductance under regular switching conditions. However, the stage may be subjected to currents above the maximum 600 A in case of a load short. For a conservative design, the maximum current the switching elements have to interrupt is maximum current they can sustain, i.e. 1.9 kA (see Sec. 4.1.1).
There exist several approaches to prevent an overvoltage condition across switching elements during turn-off. One approach is to operate the switching elements in the linear region, until the complete energy stored in the parasitic inductances is dissipated ("active clamping") [33]. In the presented design, this technology can not be implemented without sacrificing the fast current turn-off. Storing the energy transiently in snubber capacitors and dissipating it slowly after the pulse is another option. A comprehensive overview over conventional topologies is given in [64]. The solution implemented in this work suitable for very fast switching elements is depicted in Fig. 4.9a [32].

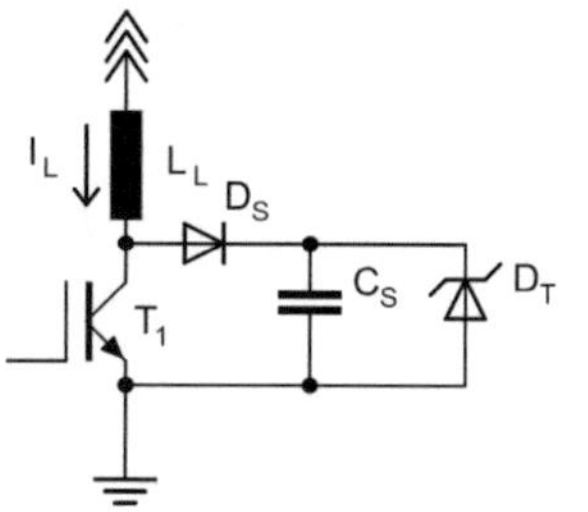

(a) The energy stored in L_L is transfered via D_S into the snubber capacitor C_S. The TVS-diode D_T prevents overcharging of the capacitor.

(b) A photo of the low-inductance snubber board to be soldered directly on the device leads of six parallel IGBTs. Each IGBT has an individual snubber diode and all snubber capacitors are connected in parallel. The bottom layer of the board is the return conductor.

Figure 4.9: Equivalent circuit diagram of the implemented snubber circuit (a) and photo of the actual arrangement (b).

The energy stored by the load current I_L in the parasitic inductance L_L is transfered after the turn-off of switching element T_1 into the snubber capacitor C_S. Diode D_S prevents a short of C_S when T_1 turns on. To prevent a significant reverse recovery current from slowing down the turn-on process of T_1, D_S

is selected to be a SiC Schottky-diode. The Transient-Voltage-Suppression (TVS) diode D_T prevents capacitor C_S from overcharging. For this design, the maximum stage current is assumed to be 1.9 kA. For measurements, the inductance of one stage was determined to 76 nH (see Sec. 4.4.2). The value of C_S is depending on the energy E_P stored in L_L, the initial charging voltage V_0 of C_S and its maximum permissible charging voltage V_S according to Eq. 4.5.

$$C_S = \frac{2 \cdot E_P}{(V_S^2 - V_0^2)} \tag{4.5}$$

D_T is chosen with a breakdown voltage slightly above the stage voltage as not to cause losses during the charging process of the stage capacitor and to allow for the maximum voltage swing across C_S. Therefore, C_S will be pre-charged to the stage's charging voltage. The maximum permissible charging voltage has to be equal to the voltage capability of the switching element, i.e. 1.2 kV. The snubber capacitance, hence, calculates to 630 nF. To enable a fast energy transfer from L_L to C_S, it is crucial to keep the circuit inductance as low as possible. Paralleling many diodes D_S, a high current can be conducted without substantial ohmic losses. The design of the snubber circuit used in this work can be seen in Fig. 4.9b. To reduce the inductance as much as possible, all the capacitors were arranged at the backside of the PCB so the current flows over the whole width of the PCB. Forward and return path of the current are on opposite sides of the board. The best possible magnetic flux compensation is achieved by minimizing the board thickness. Fabricating the board with a thickness of 0.5 mm instead of the standard 1.5 mm allows, according to Eq. 4.1, a reduction of the inductance by a factor of three. A minimal distance to the device leads is achieved by soldering the PCB directly onto the switching elements.

4.2.3 Overcurrent protection

The dynamic load behavior as discussed in Sec. 2.1.1 can lead to a complete short of the GESA device. In order to protect the switching devices from destruction by overcurrent conditions, a fast current measurement is important. For increased safety, each stage can measure the current through the switching elements and can turn-off independently of the rest of the generator or the need for an external command.

In literature, different approaches for fast and compact current measurements are presented. One way is to use the collector-emitter voltage as measured across the switching element and deduce the actual current [64]. However, the voltage drop across the device is not linear with current and dependent on the junction temperature. The very high current rise rates resulting from the use of the gate-boosting circuit demand for a very low-inductive measurement to prevent inductive signal pick-up. The geometric dimensions of a switching device and the magnitude of the voltage to be measured (1 kV in blocking stage) render a very low-inductive measurement challenging.
Another possibility is to measure the induced voltage in the parasitic inductance of the device leads to calculate the dI/dt and infer the total current by integration [75]. For an easy implementation, the authors of [75] used packages with a Kelvin-emitter for measuring the potential difference between gate and power emitter. Since the presented design relies on standard TO-247 devices, additional measurement leads would be required and increase the complexity of the design.
The influence of the magnetic field can also be measured by external means such as a Rogowski coil monolithically integrated in the PCB [92]. Due to the limited isolation distance between the device leads, no additional traces could be fitted between them.
The technology used in this work is to measure the current by means of a fast shunt resistor. Key in the design is a low-inductance approach in order to reduce an inductive component in the signal. Due to the arrangement of pulse switches and pulse capacitors (connected via broadside traces, see Fig. 4.5), also the shunt resistor can be distributed over the whole width of the PCB. Thereby many SMD components can be used in parallel, reducing the overall inductance. The resistive elements themselves are low-inductive metal-plate resistors with high pulse load capability [93]. Since the current passes in the opposite direction in the layer above the resistors, magnetic coupling reduces the inductance even further. Inductive signal pick-up in the measurement traces is reduced by routing them as close together as possible and covering the adjacent layer with copper to act as a compensation winding. Nevertheless, the shunt resistor $\mathrm{R_S}$ will incorporate a certain parasitic inductance $\mathrm{L_P}$. The fast-rising load current $\mathrm{I_L}$ will induce an additional voltage into $\mathrm{L_P}$, distorting the measurement. By adding an RC-element to the circuit ($\mathrm{R_C}$, $\mathrm{C_C}$), the inductive overshoot of the signal can be compensated. The equivalent circuit of the arrangement together with a photo of the implementation can be seen in Fig. 4.10.

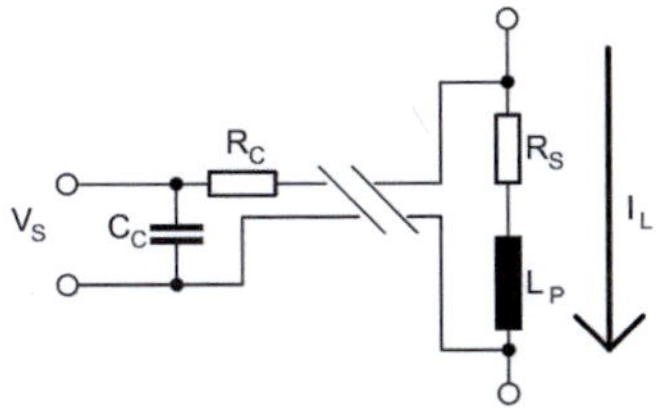

(a) The load current I_L is measured by the ohmic shunt resistor R_S, containing a parasitic inductive component L_P. By adding an RC element before measuring the voltage V_S, the inductive component can be compensated.

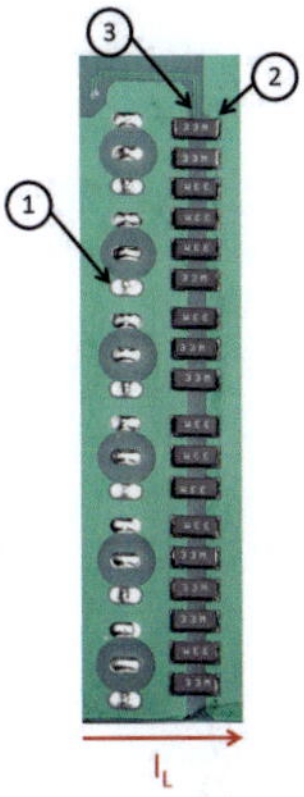

(b) The load current I_L passes through the IGBTs (1) and the shunt resistor via a broad-side trace. It consists of 18 individual metal-plate SMD resistors (2) in parallel. The measurements leads (3) are very close together and routed in parallel.

Figure 4.10: Equivalent circuit diagram of the shunt resistor (a) and photo of the implementation (b).

For complete compensation, the RC time constant must equal the LR time constant as expressed by Eq. 4.6.

$$R_C \cdot C_C = \tau = \frac{L_P}{R_S} \tag{4.6}$$

A complete compensation is advisable if a very accurate current reading is required. Increasing the time constant of the RC-element, on the other hand, also slows down the circuit's response. Instead of a complete compensation, the values of the compensation network were chosen in experiments to allow for a maximum current rise rate of 4 kA/µs. The threshold was set to 650 A (see Sec. 5.3.4). The resulting time constant τ is around 120 ns. While for the measurements presented in this chapter the overcurrent protection was deactivated to determine the maximum possible rise rate and deduce the circuit inductance, it was validated in the small-scale assembly (see Sec. 5).

4.2.4 Discharging the pulse capacitors

The high stage capacitance of 100 µF in combination with a charging voltage of 1 kV poses a considerable risk when operating the stage. In principle, shorting the capacitors after each operation by hand is possible, the high number of stages necessitates an automated shorting switch. The pneumatic shorting switch used in the final generator assembly is presented in Sec. 6.2.1. Besides the obligatory safety circuit, an additional discharge unit was implemented in each stage to provide the means to discharge the stage capacitor by software, before the hardware switch shorts the capacitors, potentially damaging the boards by the resulting high currents or voltage reversal of the pulse capacitor. As for the charging voltage distribution discussed in Sec. 2.4.1, the discharge unit uses high-voltage SMD relays as switching elements. Two dump resistors with a value of 1 kΩ each are operated in parallel to discharge the stage capacitance in less than 0.1 s (5τ). To ensure the reliable operation of the resistors, special pulse resistors were used with an energy absorption capability of 400 J each [94]. Compared to the energy stored in the stage capacitors of 100 J the safety margin is largely sufficient.

4.3 Control logic

The control logic on each stage comprises means of communicating optically with the main control unit on ground potential to exchange status parameters such as stage voltage, temperature, and the switching commands. Its main functionality is the correct switching signal generation for the pulse switches in the course of a pulse.

4.3.1 Fast and cost-efficient optical data transmission

Following the discussion of Sec. 2.3.2, the approach in this work is to use optical signal transmission for isolated communication with all 149 stages. The requirement of a synchronization of all stages within several nanoseconds requires a very fast combination of transmitter and receiver. Whereas commercial solutions for high bandwidth transmission are available, they may require a DC-balanced protocol [95]. To omit the resulting restrictions with respect to protocol, only devices capable of arbitrary signal transmission have been investigated. Due to the high amount of required transmitters/receivers, costs of individual devices and ease of handling are an important factor. Very fast photo diodes are available on the market, but the combined costs of the devices themselves and the fabrication and installation into a proper housing exceed the costs of commercial diodes already encased in an appropriate housing. For those reasons, the combination of a fast light-emitting diode (LED) in a housing (HFBR-1527Z) as transmitter and a fast PIN diode including a pre-amplifier as receiver (HFBR-2526Z) is employed in this design. The drive circuitry around them was developed in order to meet the specifications.

Driving optical transmitters

The selected transmitter HFBR-1527Z is cost-efficient, but according to the datasheet (optical rise time $t_r = 12\,\text{ns}$ and fall time $t_f = 9\,\text{ns}$ [96]) not fast enough for this project.

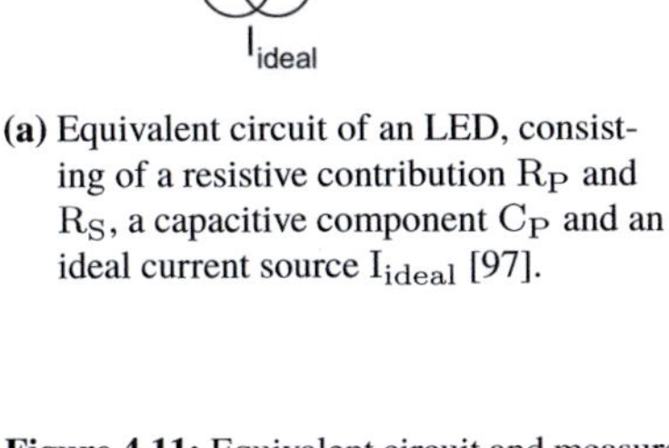

(a) Equivalent circuit of an LED, consisting of a resistive contribution R_P and R_S, a capacitive component C_P and an ideal current source I_{ideal} [97].

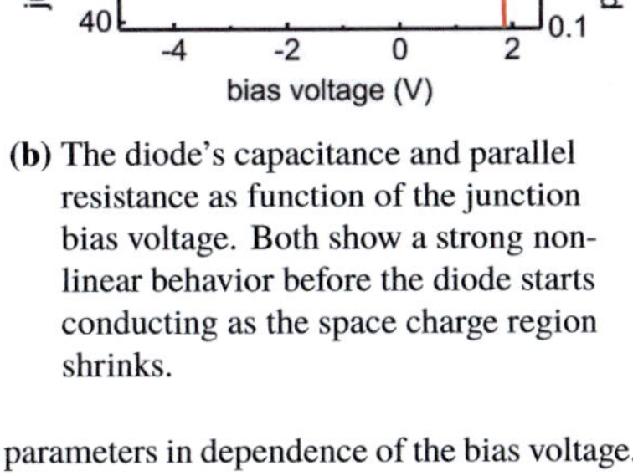

(b) The diode's capacitance and parallel resistance as function of the junction bias voltage. Both show a strong non-linear behavior before the diode starts conducting as the space charge region shrinks.

Figure 4.11: Equivalent circuit and measured diode parameters in dependence of the bias voltage.

Fig. 4.11a illustrates the equivalent circuit of an LED: under negative or small positive bias, the device behavior is dominated by its junction capacitance C_P. Its value is given by the thickness of the space-charge region within the diode. With increasing bias, the thickness of this region decreases, causing the junction capacitance to increase [97]. The measurement shown in Fig. 4.11b demonstrates the dependence of the junction capacitance on the bias voltage. As can be seen, the parallel resistance R_P drops with increasing bias until the device starts to conduct. It then behaves like an ideal diode (current source I_{ideal}) with a parasitic serial resistance. The vanished junction capacitance is replaced by a capacitive contribution due to the limited charge carrier mobility within the device [97].

The main contributions slowing the turn-on process are, hence, the charging process of the junction capacitance [98] and the limited charge carrier diffusion speed [97]. After turn-off, the charge carrier lifetime dictates the optical signal decay [97].

To foster a fast turn-on of the device, it is beneficial to bias the LED close to its knee-voltage with a small bias current [98]. Using an RC peaking network allows for a fast charging of the junction capacitance during turn-on as well as for a negative bias to extract charge carriers during turn-off. As demonstrated in [98], high drive currents shorten both optical rise and fall times.

The resulting drive circuit is shown in Fig. 4.12a. The LED D_L is driven by a fast current feedback amplifier Q_1, since they allow to adjust gain and bandwidth independently [99]. The feedback resistors R_F are split to reduce their parasitic capacitance. The bias voltage in off-state was set to 1.1 V, far enough from the knee-voltage of 1.8 V [96] as to include a safety margin for thermal drifts. In steady on-state, R_{DC} serves as current-limiting resistor. The peaking network consisting of C_{P1} and C_{P2} was adjusted empirically to the diode to deliver the best performance. The maximum current is limited by R_P to the maximum current as given in the datasheet (120 mA [96]). Fig. 4.12b illustrates the effect of different capacitance values of C_{P1} on the pulse shape. Using a small value capacitor slows down the leading edge of the pulse, whereas a high value causes a significant overshoot. A value of 220 pF was chosen, as it allows for a rise and fall time of the optical signal of approximately 3 ns without significant overshoot. The three-fold increase in device switching speed renders the diodes suitable for the use in this project, since timing jitter associated with the limited optical rise time will be below 1 ns.

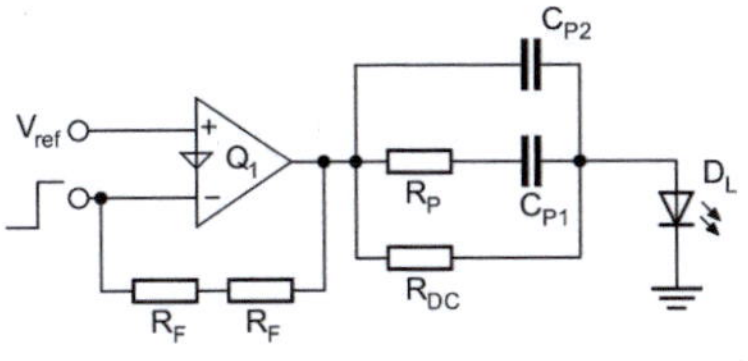

(a) The LED drive circuit used in this work. The peaking network is tailored to the specific diode to compensate its junction capacitance.

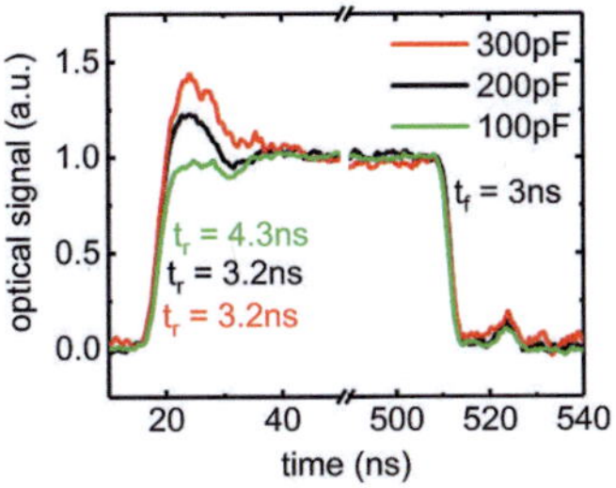

(b) Optical output signal of the employed LED for different peaking capacitors.

Figure 4.12: The developed LED drive circuit and corresponding measurements for achieving approximately 3 ns rise and fall times.

Light detection

On the receiver side of the optical connection, the selected device HFBR-2526Z is with a specified typical rise time of 3.3 ns [100] already fast enough to meet the requirements. Due to the implemented high-power LED drive circuit presented earlier, excessive optical signal has to be prevented from reaching the receiver. Especially at short fiber length below one meter, very good polishing of the optical fiber ends can lead to a saturation of the pre-amplifier. Fig. 4.13 shows a measurement of such a case. The original signal is around 11 ns shorter than the received signal. Although a high LED drive current is necessary to achieve a fast signal, the design ensures the optical signal remaining within the dynamic range of the amplifier. Additionally, the rising edge (transition light off to light on) is used for all precise timing purposes as it cannot be skewed by the saturation of the pre-amplifier. Details on the trigger signal recognition are discussed in Sec. 5.1.2.

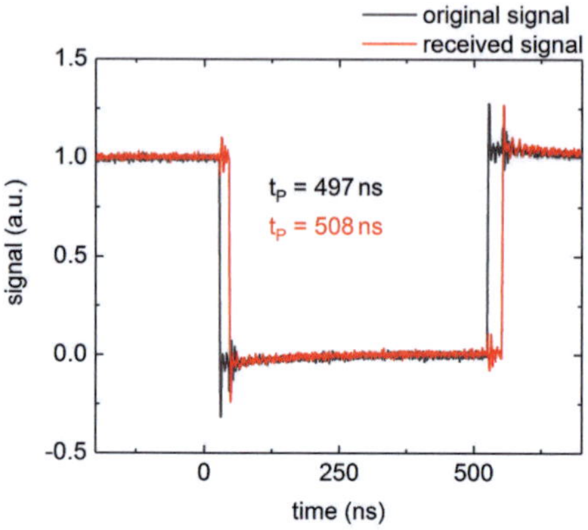

Figure 4.13: Original and received electrical signal using an LED and receiver combination with excessive optical signal. Saturation of the built-in pre-amplifier results in a pulse width distortion.

4.3.2 Switching signal generation

The new approach to generate the control signals for the stages' pulse switches locally on each stage instead of by centralized control unit (see Sec. 2.3.2) demands for a precise timing in the order of several tens of nanoseconds. On the other hand, for monitoring the stage parameter such as temperature or charging voltage, peripheral devices such as ADCs and serial ports are required. From the manufacturing point of view, only devices solderable by hand can be used for this project. Ball-grid arrays and other packages with connections underneath the device cannot be handled in the institute's workshop and, hence, would impede the development process.
Some researchers use Field Programmable Gate Arrays (FPGAs) [101] for the control of fast pulsed power generators. In principle, they can meet many of the demands also in this project. Running at high clock frequencies, the parallel computing capability allows for fast signal processing. Their main disadvantage is, however, the dependence of the gate-to-gate transit time of a signal on the performed calculation. Therefore, the transit time uncertainty introduces an additional source of timing jitter. Due to their complexity, FPGAs feature more than 100 pins rendering their manual soldering a challenge.

Since the computational tasks to be performed on each stage are not very demanding, microprocessors can accomplish many of the required tasks. They come in small packages and different peripheral features, so the most cost-efficient solution can be selected easily[8]. On the downside, precise absolute timing is challenging, since fast reaction on external events requires the use of interrupts. The time delay between the occurrence of the external event and the execution of the interrupt service routine is dependent on the processor state at the time of the external event and, hence, cannot be predicted with sufficient accuracy. Precise relative timing, however, can be achieved using the Timer-Counter (TC) peripheral device. It allows for setting or resetting certain output pins depending on the number of passed clock cycles, without intervention of the CPU. To cope with this limitation, a combination of microprocessor and Complex Programmable Logic Device (CPLD) is used [102]. CPLDs are devices with programmable parallel logic, incorporating more than 1000 logic gates and having a constant pin-to-pin delay independent on performed calculation [103]. Moreover, the selected device (MAX3000, Intel) offers 5 V capabilities, facilitating the interface with standard TTL logic.
In the presented project, only a very accurate synchronization of the leading edge is important. Therefore, the microprocessor handles all communication, data acquisition and "slow" switching signal generation. To create a synchronous leading edge with all stages, the CPLD is used to recognize the pulse start command and to execute the first turn-on of the pulse switch. Afterwards, the TC module controls the pulse switch with precise relative timing. The microprocessor also supplies the clock signal to the CPLD. Both microprocessor and CPLD run at a clock frequency of 100 MHz, giving rise to a time uncertainty of 10 ns. The experiments in a small-scale generator setup (see Sec. 5.3.1) could validate the domination of this timing jitter in the total jitter of the pulse execution of several stages.

4.3.3 EMI testing

In the final generator arrangement, each stage will be subjected to a substantial amount of electromagnetic interference. Especially the high-speed logic circuitry used on each stage is vulnerable and has to be shielded accordingly. To

[8] This project uses: Microchip ATSAM4N8A, 4€ [59]

test the designed shielding enclosure, an EMI test bed was set up to simulate the conditions a stage will experience in the final assembly. Fig. 4.14 illustrates the used test setup.

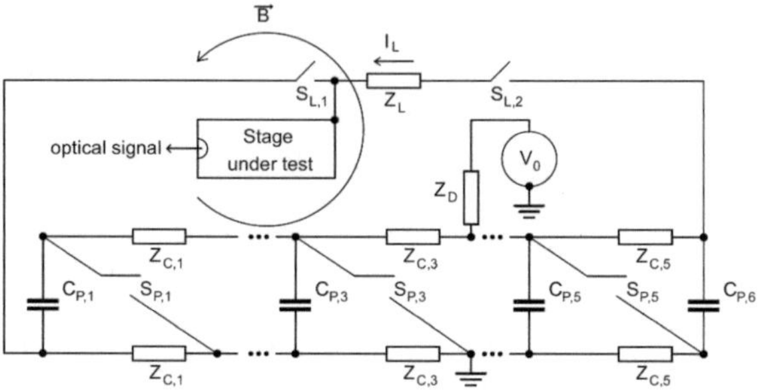

Figure 4.14: Circuit diagram of the employed spark-gap based Marx generator used to simulate both capacitive and magnetic coupling. The stage-under-test is connected to the output of the generator.

It consists mainly of a traditional spark-gap based Marx generator with six stages, each stage with a pulse capacitor of approximately 100 nF and decoupling inductors of 380 µH [104]. The capacitors were charged to $V_0 = 35\,\text{kV}$. By grounding the generator at the center, the isolation distance to the surrounding is reduced. The generator was connected to a load impedance of 100 Ω and several microhenry of circuit inductance. Connecting the stage-under-test to one side of the load impedance, the fast shift in the stage's potential with respect to ground once the generator erected was used to simulate the capacitive coupling in the final arrangement. For simulating the magnetic coupling, the load current was lead close by the vulnerable circuitry. In addition, the employed spark-gaps feature a broad-band emission spectrum [105] that was used to test the robustness against radiated EMI. During the experiments, the stage was powered using a battery and a DC-DC converter from 12 V battery voltage to –500 V supply voltage (see Sec. 2.4.2). The state of all input pins, including the clock signal, was monitored by means of an optical fiber. For this experiment, the CPLD was programmed to output the clock signal divided by four, only if neither a signal was received on either input pin, nor any optical signal was received. In case the clock signal could not be recorded over the optical fiber, hence, electromagnetic interference caused an incorrect signal recognition.

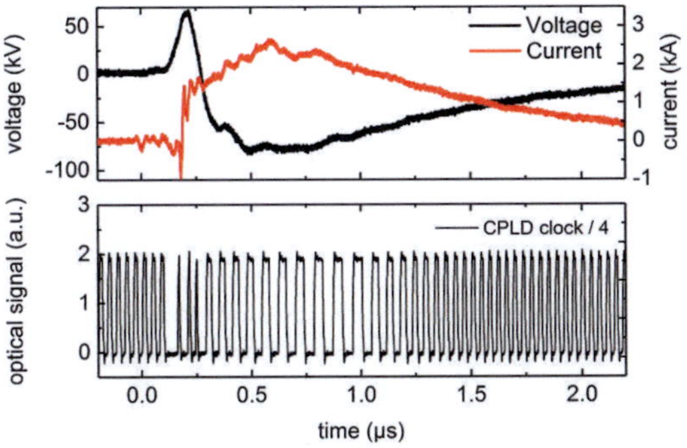

Figure 4.15: Upper graph: Output current (red trace) and output voltage (black trace) of the conventional Marx-generator to which the stage-under-test was connected.
Lower graph: Transmitted divided clock signal.
When the Marx-generator is erecting, a false optical recognition occurs as there is no more signal from the stage. Afterwards the clock signal returns with varying speed, indicating a clock generator failure.

Fig. 4.15 shows a test with an unshielded board. The upper graph displays output voltage and load current of the Marx generator. Due to the grounding point, the output voltage of the generator is negative with respect to ground and only half of the sum charging voltage. The negative voltage of −100 kV is approximately equal to the design output voltage of the semiconductor-based generator. The positive peak in the beginning of the pulse is a result of the positive half of the Marx generator erecting before the negative half – increasing the voltage swing the stage is subjected to. The load current crests at around 2.5 kA, exceeding the expected current of 600 A by a factor of four. With a look on the optical signal transmitted from the stage (lower graph), it is obvious that EMI influences the circuitry. Upon ignition of the generator, the clock signal fails – indicating a wrong signal reception. Afterwards, the clock signal returns but the clock speed is first slowing down, then speeding up. This behavior can only be explained with interference into the clock signal generator. Since precise timing is key in this application, both types of failures have to be avoided.

For EMI protection, the whole PCB was inserted into a copper box of 1 mm thickness. The box was closed on five sides and a copper block was mounted onto the PCB to seal the sixth side. Good electrical contact between the box and the block was achieved using $CuBe_2$ contact springs. The arrangement is illustrated in Fig. 4.16.

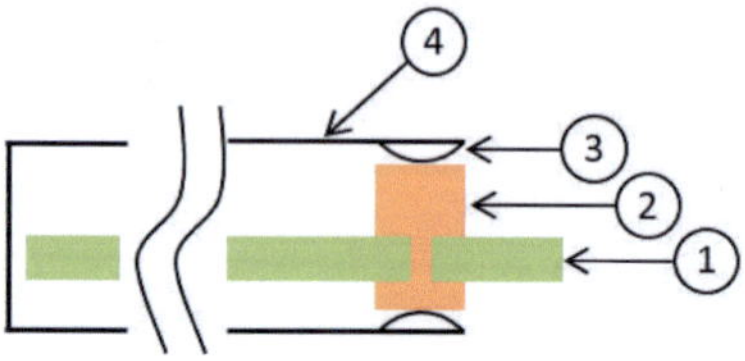

Figure 4.16: Simplified illustration of the shielding enclosure. The PCB (1) is inserted into a copper box (4). The opening is sealed by means of a copper block (2), mounted onto the PCB and contact springs (3) connecting the block to the box.

In Fig. 4.17a, the stage is depicted in the designed shielding enclosure. When the box is subjected to a varying magnetic field, compensating currents are induced depending on the orientation of the magnetic field. Since the box is closed on all sides (and the copper block has connections from one side of the PCB to the other), the compensating currents are not limited to single faces of the enclosure but can flow over the whole surface. Signal filters were used to block voltage transients on all logic traces leaving the shielding enclosure. The pulse switches are within the shielding enclosure, as the fast gate signals cannot be shielded effectively. The main source of EMI in the housing are voltage transients in the order of 1 kV/10 ns resulting from the switching elements within the housing, but did prove not to influence the circuitry.

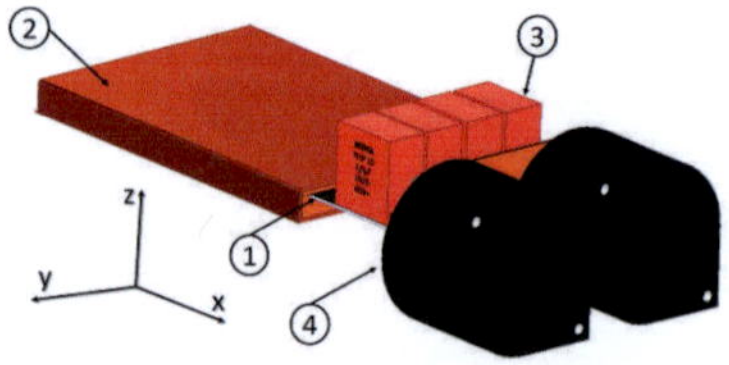

(a) The shielded stage corresponding to the measurement in Fig. 4.18.

(b) The fast logic circuitry is shielded by means of a tinplate SMD shielding cabinet (photo without cover).

Figure 4.17: For adequate shielding, a two-layer approach was chosen. The outer shielding made from copper (a) and the inner shielding made from tinplate (b).

As mentioned in Sec. 2.3.4, the combination of high-conductivity shielding material and high-permeability material is advisable for very sensitive parts. For this reason, the fast logic circuitry was located under a SMD shielding cabinet made from tinplate [106] with an attenuation of 60 dB in the range of 500 MHz to 3 GHz where copper is not as effective as in the low-frequency range. A photo of the arrangement without the top cover of the SMD shielding cabinet is shown in Fig. 4.17b. In addition to shielding against magnetic interference, the closed compartments (both box and SMD cabinet) work well against capacitive coupling. With the presented shielding, the stage was again tested in the EMI test bed. The results are presented in Fig. 4.18.

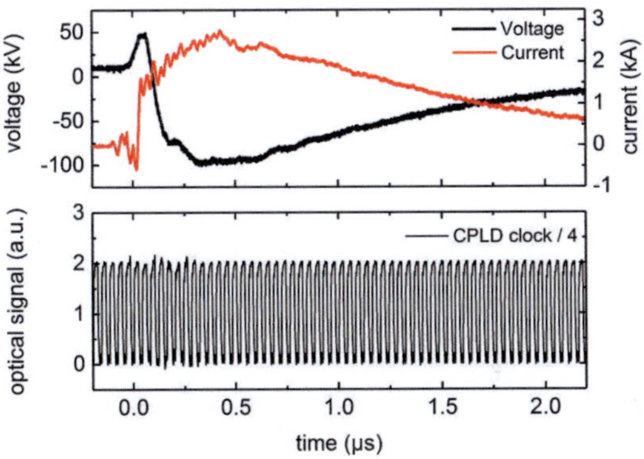

Figure 4.18: Upper graph: Output current (red trace) and output voltage (black trace) of the conventional Marx-generator to which the stage-under-test was connected.
Lower graph: Transmitted divided clock signal of one stage with proper shielding. As can be seen, the transmitted signal is not influenced by the external Marx-generator, proving the efficiency of the shielding.

As can be seen, the signal as transmitted by the stage does not show any irregularities once the generator erects. The efficiency was validated for several shots. Moreover, the path of the load current was changed in orientation. Whereas the exact disturbance one stage is experiencing in a specific generator arrangement cannot be simulated without a comparable generator, these measurements can be used to estimate the necessary means for shielding the circuitry. While current- and voltage amplitude and rise time are comparable to the final experiment, mainly the duration of the pulse differs significantly from what the stage will experience in the final arrangement. The longer pulse duration may cause a saturation of the tinplate shielding cabinet. However, since the pulse duration mainly influencing the low-frequency content of the interference spectrum where the copper shielding (non-saturable) is the most efficient, this effect is expected to be of minor importance.

4.4 Design validation

4.4.1 Pulse circuit qualification

For benchmarking the stage's switching behavior, it was operated in a low-inductive test circuit. The circuit is displayed in Fig. 4.19: the stage (as described in this section), consisting of the pulse switch T_1, the pulse capacitor C_1 and the freewheeling diode D_1 is connected to a variable load resistors R_L. The switching commands are generated by the control logic on the stage. In the following measurements, the load current is plotted as negative due to the negative polarity of the stage's output voltage. In addition to the discrete elements as shown in the circuit diagram (Fig. 4.19), each element as well as all connections and leads have a certain stray inductance. For the development process of the generator, knowledge of the stray parameters is very important. The stray inductance can be estimated by comparing the differences in voltage and current rise time.

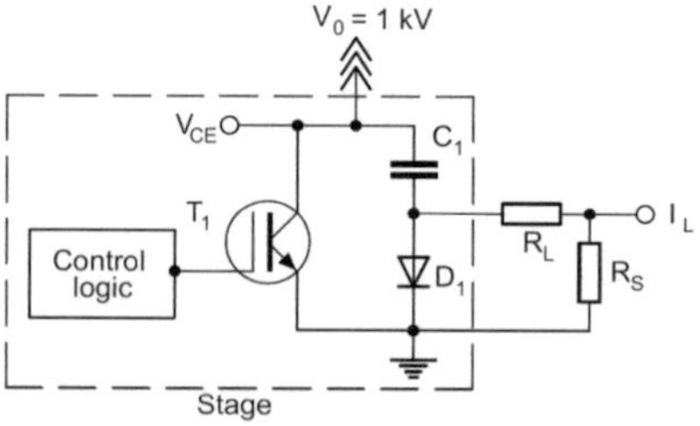

Figure 4.19: One stage (switching element T_1, stage capacitor C_1, and free-wheeling diode D_1) in a low-inductive pulse circuit with load R_L. The collector-emitter voltage V_{CE} is measured directly across the switching element, the load current I_L is measured by means of a shunt resistor R_S ($R_S << R_L$).

To this end, one stage was operated under varying load condition at 1 kV charging voltage. Fig. 4.20 shows a typical turn-on event of a stage with two different load resistors. The shape of the collector-emitter voltage (red) does not change significantly in dependence of the load resistor. However, the load current rises fast for a larger load resistor of 2.5 Ω than it does for a smaller load resistor of 1.66 Ω. For a better comparison, the load current is plotted normalized to its maximum value of 400 A and 600 A, respectively.

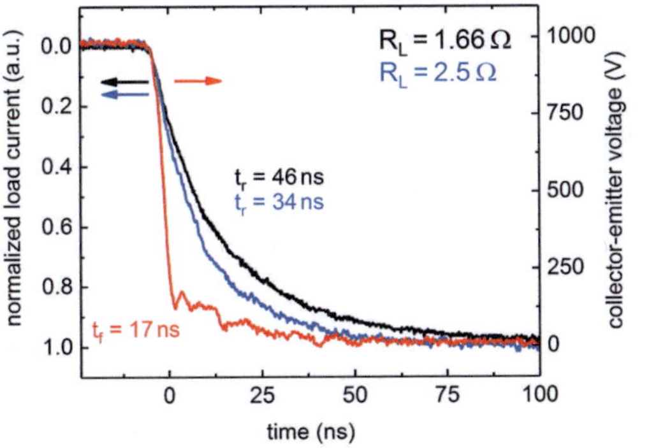

Figure 4.20: Collector-emitter voltage and normalized pulse current for 2.5 Ω and 1.66 Ω load. The rise time is mainly determined by the circuit inductance.

As discussed in Sec. 2.5, the total rise time of a combination of two circuits with finite rise time can be calculated as the geometric sum of the two. Here, the switch has a limited rise time as can be measured by the voltage fall time across it (τ_V). In addition, the connected LR circuit reacts with its time constant (τ_{LR}, compare Eq. 4.6). The measured current rise time through the load resistor can be considered the total rise τ_I. Therefore, the parasitic inductance can be deduced according to Eq. 4.7.

$$L = R_L \cdot \tau_{LR} = R_R \cdot \sqrt{\tau_I^2 - \tau_V^2} \tag{4.7}$$

The calculation parameters and resulting stray inductance for the two traces plotted in Fig. 4.20 are displayed in table 4.1.

Table 4.1: Calculated stray inductance values.

τ_V	Load resistor (peak current)	τ_I	Calc. circuit inductance
17 ns	2.5 Ω (400 A)	34 ns	72 nH
17 ns	1.66 Ω (600 A)	46 ns	71 nH

Both values are identical within the resolution of the measurement. As a worst-case estimation, the total stray inductance of one stage within the generator can be assumed to be 72 nH not including the stage-to-stage wiring. Considering the circuit diagram in Fig. 4.19, the majority of the inductance can be either located in the pulse circuit (C_1, T_1) or the load circuit (R_L, R_S).

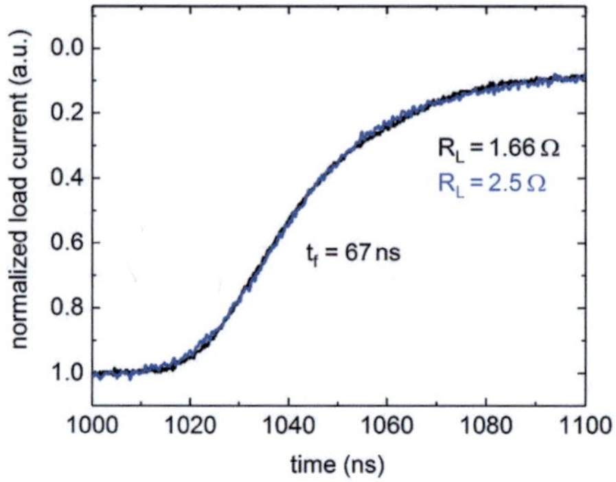

Figure 4.21: Normalized pulse current for 2.5 Ω and 1.66 Ω load, trailing edge of one pulse. Since the current traces for both currents coincide, the turn-off speed is limited by the switching devices.

By comparing measurements of the turn-off characteristic of the stage (see Fig. 4.21) for different load resistors, the location can be determined. During turn-off, any stray inductance in the load circuit would drive a current through the free-wheeling diode D_1 when the switching element is turned off. The decay of this current is governed by the ratio of stray inductance and load resistance (neglecting the voltage drop over the free-wheeling diode) and, hence, should differ for different load resistors. The graph shows normalized load current traces for 1.66 Ω and 2.5 Ω load resistors, both coinciding within the resolution of the measurement with a fall time of 67 ns. Therefore, majority of the stray inductance is located in the pulse circuit instead of the load circuit.

4.4.2 Fast current rise time measurement

The rise time of the generator will be determined by the combination of stage rise time, generator inductance and stray capacitance to ground (see Sec. 2.3). To allow for a sufficient safety margin with respect to the stage switching speed, the figure of merit used in the design of one stage was the current rise time at maximum load current. Limiting the energy dumped in the solid carbon load resistors, the pulse length was kept around several microseconds only in the experiments.

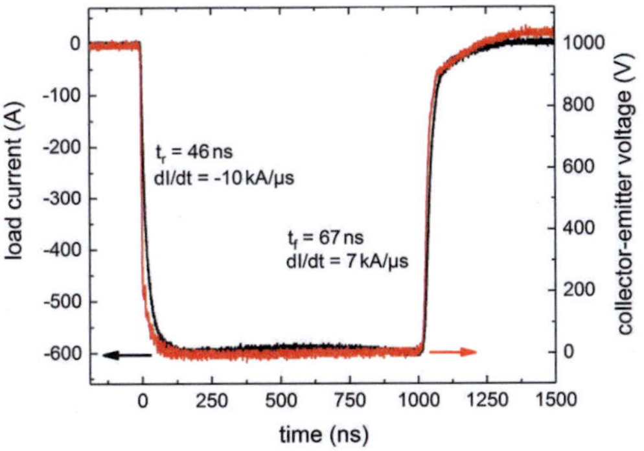

Figure 4.22: Load current (black trace) and collector-emitter voltage (red trace) across the switching element of the stage at nominal charging voltage of 1 kV. Connected to 1.66 Ω load resistance, a current rise time of 46 ns can be achieved for 600 A load current.

Fig. 4.22 displays current and voltage of one stage operated in the pulse circuit presented in the previous section. On a load resistor of 1.66 Ω, a current rise time of 46 ns can be achieved with a load current of 600 A. This equals a current rise rate of around 10 kA/µs using six IGBTs in parallel. Compared to the achievable 4.2 kA/µs current rise rate with a single device (see Sec. 3.3.4), it becomes clear that the switching time of the IGBTs is much faster than the response of the pulse circuit. The falling edge, being of minor importance to this project, is with a transition time of 67 ns 30 % slower than the rising edge. A current fall rate of 7 kA/µs can be achieved. The current fall rate is roughly equal to six times the current fall rate for a single device operated at 100 A (compare Fig. 3.26).

4.4.3 Arbitrary switching sequence

In addition to a fast rise rate, also the capability of performing fast and arbitrary switching sequences is a requirement for the designed stage. One stage was operated with 4 Ω load, delivering a nominal pulse of 250 A, 1 kV with 500 kHz repetition rate at 50 % duty cycle. Due to the limited energy absorption capability of the employed resistors, the stage was operated in burst-mode only. Besides the demonstrated waveform, any on/off sequence within 100 µs and toggle frequency below 500 kHz can be performed by the control logic. The resolution of the employed counter is, limited by the clock speed of the TC module of the microprocessor, 20 ns.

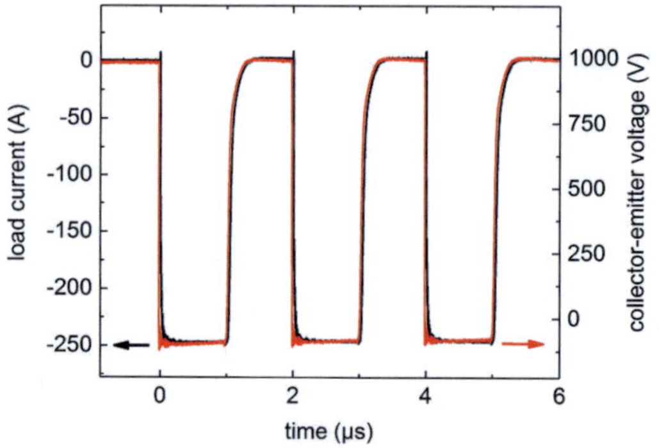

Figure 4.23: Load current (black trace) and collector-emitter voltage (red trace) across the switching element of the stage at nominal charging voltage of 1 kV. Connected to 4 Ω load resistance, arbitrary switching sequence generation is demonstrated by a 500 kHz waveform with 50 % duty cycle.

5 Proof-of-concept assembly

Prior to the construction of the full-scale version, the operation of several stages in a generator arrangement had to be tested. Their interaction and communication via the developed optical bus system together with the generator's main control unit is described in this section. All crucial functionalities such as a low-jitter triggering of many stages, a fast output rise time and arbitrary output waveforms at 600 A load current are successfully demonstrated. Major results of this work presented in the following chapter have already been published as [23].

5.1 The Marx module

Resulting from the implementation of the optical bus system, several stages are logically grouped together to form a module within the generator. All stages within one module are addressed using one optical link.

5.1.1 One stage equivalent circuit

To understand the functional interaction of many individual stages as presented in the previous chapter, a glance on the circuit diagram as depicted in Fig. 5.1 is helpful. It combines the main features of stage n, having three connections to the lower stage (i.e. closer to ground) and three to the upper stage (i.e. closer to the output terminal). The charging voltage path, as described in Sec. 2.4.1, is transiently isolated using one high-voltage relay contact S_1. The same coil operates a second contact S_2 used for shorting the free-wheeling diode D_n. Hence, the pulse capacitor C_n is charged via S_1, charging diode D_C and S_2 – the return current using the pulse current connection.

The negative voltage of –500 V used as supply voltage is distributed via the diode string D_{S2} and the charging diode D_{S1} to charge the buffer capacitor C_S.

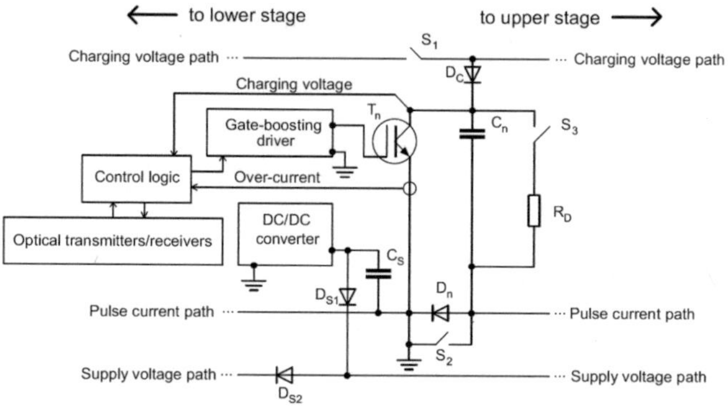

Figure 5.1: Main circuit components of one stage. The ground symbol is the local reference potential for the control logic circuitry.

The stage uses several DC/DC converters (5 V, 15 V and 80 V) to generate the required supply voltages. The pulse capacitor can be discharged by software closing the high-voltage relay contact S_3, thereby discharging C_n via R_D. To achieve fast switching, the pulse switches T_n are operated using a gate-boosting driver circuit (see Sec. 3.3). A fast shunt resistor is used to provide an overcurrent signal to the control logic to block the switching element in case of a malfunction, whereas the charging voltage across the pulse capacitor can be measured by the control logic. Arbitrary switching sequences with precise timing are realized by pre-programming the control logic as discussed in Sec. 4.3 via optical fibers. Using a novel optical bus system for interconnecting the stages, a synchronization within 10 ns and a fast pulse interrupting in case of an overcurrent detection can be achieved.

5.1.2 Combination of stages via an optical bus system

The optical bus system investigated in this work allows for a reduced complexity of the control circuitry as compared to a direct link from the control circuitry to each individual stage. Fig. 5.2 illustrates the concept: a PC serves as user interface and transmits the respective commands to a control unit via an optical USB interface enabling an RS232-like communication. The control unit has no direct link to each stage – instead the stages are grouped in modules. The control unit only communicates with the middle stage of each module, from where the signal is forwarded upwards and downwards. This allows for addressing one specific stage (by protocol) or for addressing all stages simultaneously. The bus system is bidirectional, so also data from one stage to the PC can be transferred. While it is in principle possible to operate all stages of the generator in one single module, the transit time of the signal from the signal infeed to the last stage limits the maximum turn-off speed in case of a failure.

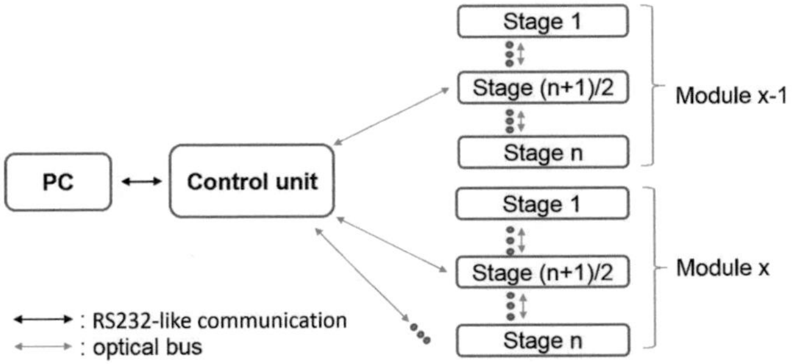

Figure 5.2: Generator control schematic, consisting of a PC as user interface, the control unit responsible for addressing x individual modules and the optical bus connecting n stages within one module.

Using the CPLDs on each stage, the routing of the optical signal can be changed depending on the generator state. Fig. 5.3 shows a connection of one branch in one module. Solid lines indicate logic connections on the stage, dotted lines symbolize a fiber optic link. Prior to pulse execution (a), the signal originating from the main control unit is forwarded to all stages – their receivers and transmitters are arranged in parallel. Therefore, each stage listens to the incoming data stream and can react when addressed. Collisions on the return data line are prevented by software. If a pulse is to be executed, the main control unit transmits the command for each stage to wait for the synchronization signal. Then, all transmitters and receivers are connected in a series configuration. For assuring all stages are ready, the control unit turns on its transmitters permanently. If a stage is ready and has not detected any error, it will forward the incoming signal. The last stage in one branch is programmed to "reflect" the incoming signal, that is to say transmit downwards what is received. The middle stage communicating with the control unit will output a logic AND connection of the signal received from above and below. Therefore, the task of checking if all stages in one module are ready for the pulse is reduced to check if light is emitted from the middle stages. Each stage can interrupt the pulse in case of a detected error by turning off its transmitters. Neighboring stages will detect the absence of the signal and start to turn off even before the main control unit has to react. Fault detection by the main control unit will lead to the turn-off of all transmitters and, hence, the generator.

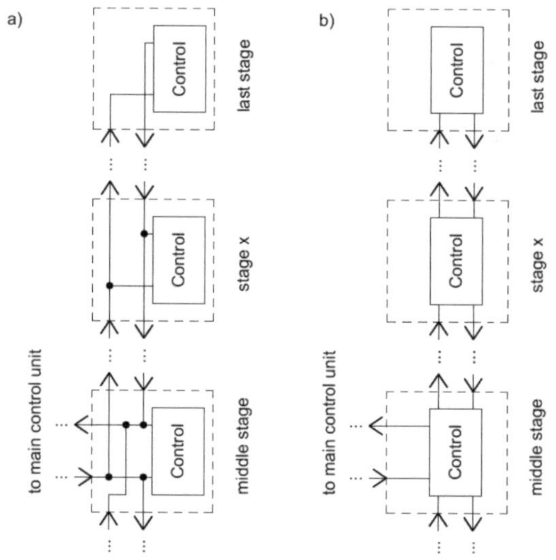

Figure 5.3: Employing a CPLD, the hardware configuration of the optical signal routing can be changed from a parallel connection of all stages during communication (a) and a serial connection for fault detection during the pulse (b).

For a good synchronization of all stages, the transit time of the signal is important. Fig. 5.4 shows a signal as measured on four consecutive receivers in a module. As can be seen, each stage in the bus introduces an additional time delay of 20 ns. The measurement was taken with an optical fiber of 1 m and therefore includes around 5 ns delay due to the transit time of the light through the fiber. Being a multiple of the clock period (10 ns), however, it can be easily compensated. Limiting the delay time between a signal from the control unit to the highest stage to below 100 ns, in the presented work a total of nine stages was used per module. As synchronization signal, a 30 ns "light-off" pulse is used – the pulse is started synchronized on its trailing "light-on" edge.

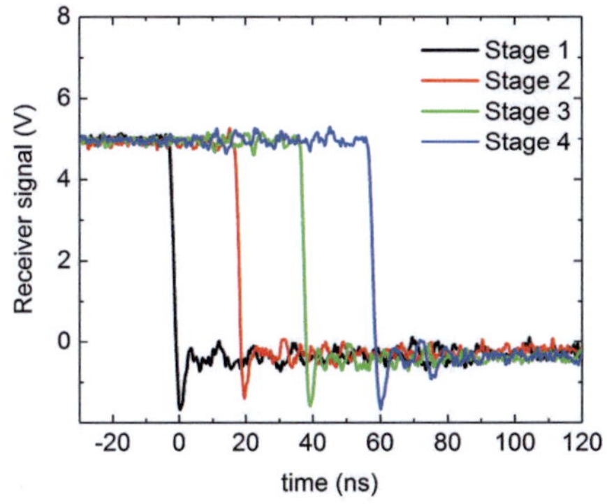

Figure 5.4: Electrical bus signal transfered from one stage to the next. Each stage contributes 20 ns time delay to the signal.

5.1.3 Software control

The functionality of the logic circuitry on each stage is distributed between the CPLD and the microprocessor. As simplification, a flowchart of the implemented state machine is depicted in Fig. 5.5. Dashed boxes indicate a state that can only hold for a certain time. The effect of each state on the charging and discharging relays (S_1, S_2, S_3, compare Fig. 5.1), the transmitter/receiver arrangement (compare Fig. 5.3) and the gate driver is summarized in table 5.1. After restart, the stage is in "idle" state allowing for communication with the user via the control unit. The stage is powered by the supply voltage and all relays are closed. The pulse capacitor is being discharged via the discharge switch. "Charging" refers to the same state, now with the discharge switch open.

This is the default state also after pulse execution as not to discharge the capacitors completely after one pulse.

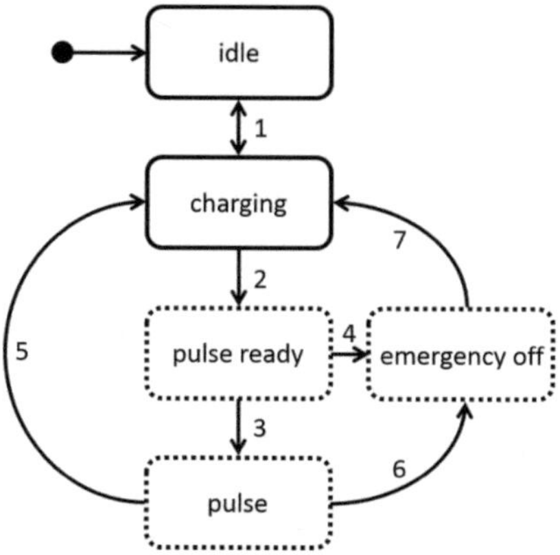

Figure 5.5: Flowchart representation of the state machine controlling one stage.

Transition (1) can be initiated in both directions via software command. From "charging" the stage can be transfered to "pulse ready" (2) via a software command. The relays open and isolate the stage, now being powered purely from the buffer capacitors. Connecting the transmitters/receivers in series configuration allows for monitoring the "pulse ready" state in all stages.

Table 5.1: Effects of the logic state on the stage's components.

State	S_1, S_2	S_3	Transmitter/receiver arrangement	Gate-driver
idle	closed	closed	parallel	disabled
charging	closed	open	parallel	disabled
pulse ready	open	open	serial	disabled
pulse	open	open	serial	enabled
emerg. off	open	open	disabled	disabled

If the synchronization signal is received within a certain time frame (3), the stage transits into "pulse" state. The IGBT gate-drivers are active and the microprocessor runs the pre-programmed switching sequence. After the sequence has ended, the stage resets into charging state (5). In case the synchronization signal has not been received within the permitted time frame (4), the stage transits into "emergency off". The same happens if in "pulse" the absence of the optical signal is detected or an overcurrent is measured (6). In "emergency off", the stage deactivates its transmitters to signal the error to adjacent stages and waits for a certain time before resetting to "charging".

5.2 The main control unit

5.2.1 Interfacing the generator to the experiment's control

The generator's main control unit represents the interface between the generator, its power supplies, the GESA experiment's control and the user (via a PC).

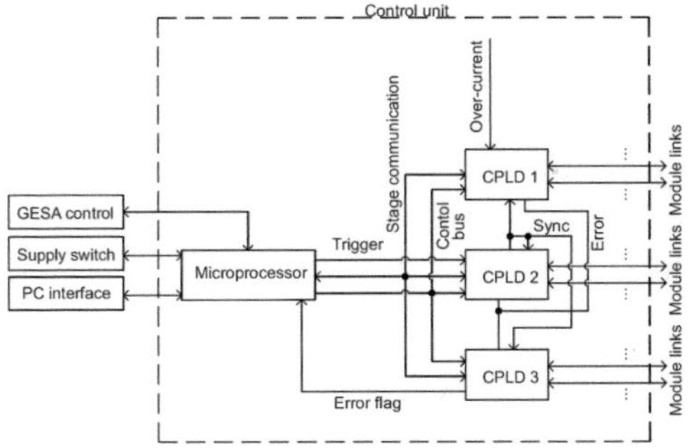

Figure 5.6: Circuit components of the control circuitry, interfacing the generator, the PC user interface, and the GESA controls.

Its main component is a microprocessor (see Fig. 5.6), controlling the supply switch disconnecting the generator from the power supplies during the pulse. Using bidirectional communication, pulse execution is prevented in case of a malfunction of the switch. The interlocks of the power supplies are controlled by the GESA experiment control. To facilitate the integration of the generator into the experiment, the interface is nearly identical to the previously used interface: the GESA control receives a "read to charge" command when the generator has been programmed successfully and sends an "execute pulse" command when charging is complete and a pulse is to be executed. Additionally, the microprocessor is used for processing the serial data stream from the PC. If a received command is intended for the generator, the microprocessor forwards it to all modules simultaneously. Due to the limited serial communication ports on cost-efficient microprocessors, again a combination with CPLDs is used. One serial port is, therefore, sufficient to communicate with the generator: the transmit port drives all available transmitters in parallel, all receivers are combined on the receive port using a logical OR connection. Overlapping telegrams are prevented by software. A control bus is used to configure the CPLDs depending on the current generator state (for details on the software, see Sec. 5.2.2). The limited size of the CPLDs, both in terms of pins and macrocells, necessitates the use of three individual devices. In order to avoid timing problems, they all use the same clock signal as provided by the microprocessors. All traces on the PCB are matched in length within 1 cm inbetween the devices and from the devices to the periphery. A fast turn-off of the generator is assured by an error signal that can be driven by each device. Besides the common functionalities described in the following section, each device has one special function. CPLD1 is used to monitor the overcurrent signal as recorded on the control unit. Its threshold can be adjusted by the PC and the measurement takes place in the main current monitor in the generator itself. CPLD2 has the functionality of creating the synchronization signal for the generator upon the reception of a trigger impulse from the microprocessor. CPLD3 can raise an error flag to the microprocessor if any error has occurred during the pulse.

5.2.2 Implementation of the main functionalities

The working principle of the main control unit can be understood with a glance on the simplified flowchart representation of implemented state machine in combination with the effect of each state on the output pins (summarized in table 5.2). All dashed boxes indicate non-persistent states. Starting in "idle" state, the control unit signals to the GESA control to disable the high-voltage power supply (HVPS) used for charging the pulse capacitors. In this state, the supply switch is closed, connecting the auxiliary power supply to the generator. All transmitters are configured to relay the data stream from the PC to the generator. When all stages are programmed, the user can set the control unit into "pulse ready" (1) which, in turn, signals to the GESA control to activate the HVPS and charge the capacitors. By sending the trigger command (2), the GESA control can initiate the pulse execution. In "prepare pulse", the supply switch is opened, isolating the generator from the HVPS and the auxiliary power supply. Once the current through the supply switch ceases, the command to open the high-voltage relays (stage state "pulse ready", see table 5.1) is transmitted to the stages before turning on all transmitter permanently. As discussed in the previous section, by comparing sent and received signals from all modules the correct reception of the command by all stages can be verified. If all stages are ready (4), the control unit transfers to the "execute pulse" state, enabling the transmission of the synchronization signal. Consecutively, all stages preform their pre-programmed switching sequence according to the stored data. After the maximum pulse length is reached (plus the time necessary to close all relays), the control unit resets to "idle" (5). In case not all modules report "ready" (6) or transmitters are disabled during the course of the pulse (7), the control unit transfers to "emergency off". Here, it disables all transmitters, preventing either the transmission of the synchronization signal (if the pulse was not started yet) or turns off the generator (in case the pulse was already started).

Table 5.2: Control unit states and their effect on output pins.

State	HVPS	Supply switch	Transmitters	Sync. signal
idle	off	closed	PC communication	disabled
pulse ready	on	closed	PC communication	disabled
prepare pulse	on	open	on	disabled
execute pulse	on	open	on	enabled
emerg. off	on	open	off	disabled

A timing diagram of the pulse execution can be seen in Fig. 5.8. The trigger signal is transmitted from the GESA control to the main control unit (falling edge). Immediately afterwards, the supply switch disconnects the power supplies from the generator (falling edge) so the stage relays do not have to break the supply current.

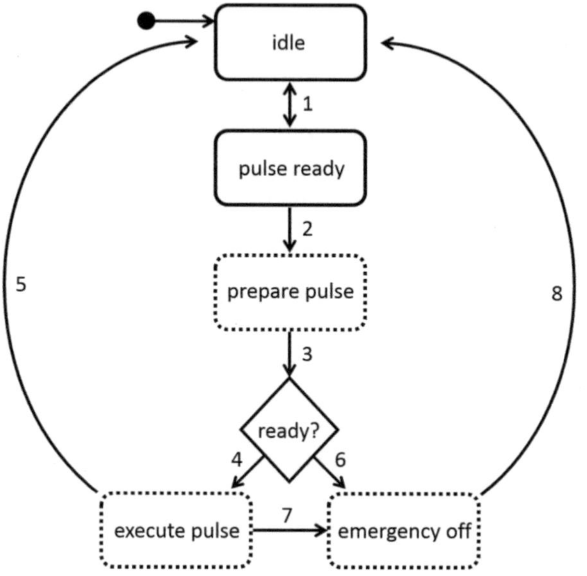

Figure 5.7: Flowchart representation of the state machine implemented in the control unit.

Once the supply switch is open, the stage opens its relays (falling edge). Since the release time of the relays is specified with several milliseconds, the pulse is executed with some milliseconds delay. The pulse is measured by means of a gate signal on one stage (spike). Once the pulse is finished, each stage waits for another five milliseconds before closing the relays again (rising edge). To ensure all relays are closed before the power supplies are connected to the generator, the control unit waits for 15 ms after the pulse before reconnecting the generator to the power supplies (rising edge). During the time the supply switch is open, each stage is powered purely from the buffer capacitor on each stage. Fig. 5.9 shows a picture of the control unit used for the 149-stage full version of the generator. By using the optical bus communication, only 17 fiber optic transmitters/receivers are necessary to communicate with all stages. In order to protect the sensitive circuitry, the PCB is enclosed in a solid aluminum housing with a wall thickness of 2 mm. To allow for an efficient shielding while being able to feed the optical fibers into the housing, solid aluminum pipes are welded to the housing forming waveguides.

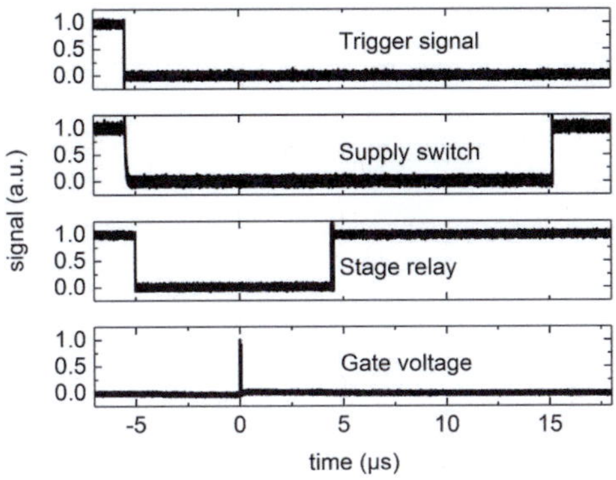

Figure 5.8: Timing diagram of the trigger signal (from the main experiment control), the opening of the supply switch (i.e. high-low transition), the opening of the stage relay, and the gate signal (i.e. pulse execution) of one stage.

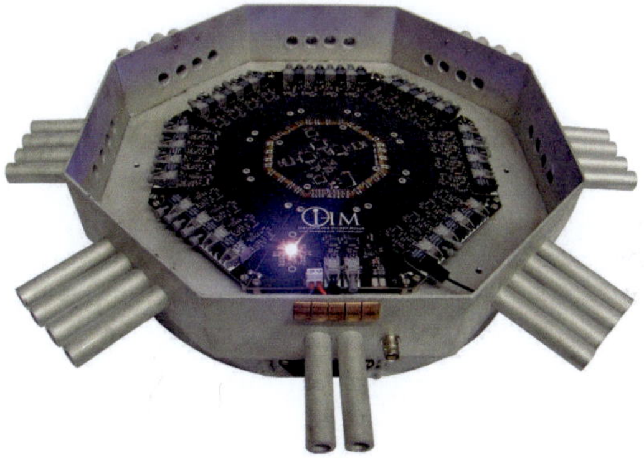

Figure 5.9: The generator's main control unit in its shielding enclosure.

Following the discussion in Sec. 2.3.4, their length to diameter ratio is above five. For easy access, the enclosure is covered with a solid lid – the sides showing a significant overlap. A good electrical contact between lid and housing is achieved by means of a flexible highly conductive $CuBe_2$ mesh. The arrangement is depicted in Fig. 5.10.

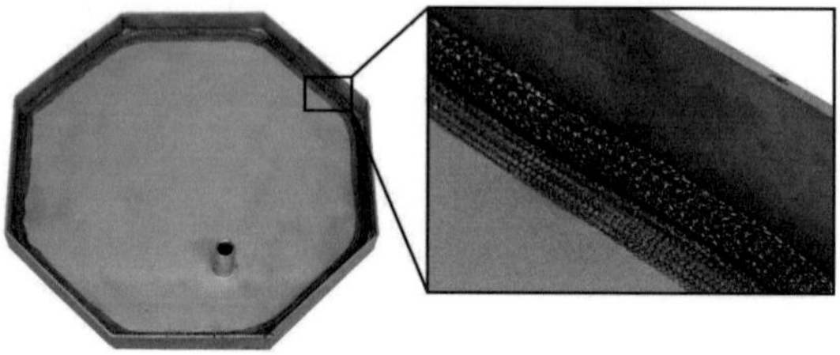

Figure 5.10: To shield the sensitive electronics within, the control unit's housing is closed by a tight-fitting solid cover, the slits between housing and cover are sealed by a flexible $CuBe_2$ mesh.

5.3 Design validation

5.3.1 Low-jitter triggering

Besides the fast switching elements, low-jitter triggering of each individual stage is key for a fast rise time. As discussed before, all stages are ready when a permanent "light on" signal is received. The synchronization signal to start the pulse execution is a 30 ns "light off" pulse created by the main control circuitry. Once this signal is recognized, the pulse is started. Using this encoding has two major advantages: primarily, using "light on" as ready signal prevents pulse execution in case of transmission errors due to bent/broken optical fibers. Secondly, using a "light off" pulse means the triggering is executed on a "light on" edge that cannot be skewed by saturation of the optical pre-amplifier (see Sec. 4.3.1). Operating four stages in one optical bus system with the control unit, the gate signal of each stage was measured. The results presented in Fig. 5.11 overlap 100 individual pulses (a total of 400 gate signals). By compensating the 20 ns time delay introduced by each stage, the total measured jitter is below 11 ns. This value is expected and caused by the limited clock speed of 100 MHz resulting in a 10 ns time uncertainty for the trigger signal recognition.

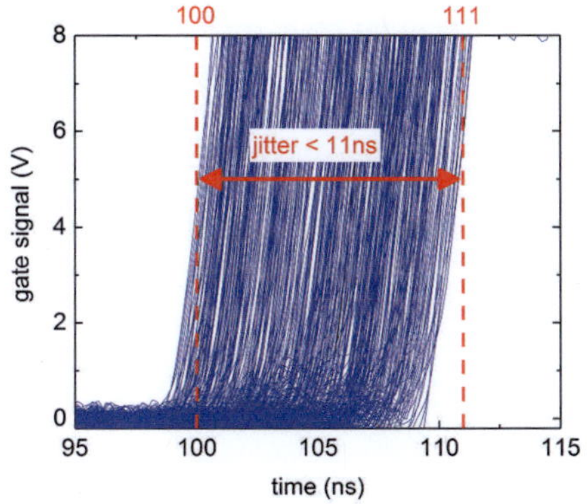

Figure 5.11: Overlap of the gate signal at the beginning of one pulse measured at four stages simultaneously for 100 pulses. The resulting jitter is below 11 ns and is mainly determined by the limited clock speed (100 MHz) of the CPLD responsible for trigger signal recognition.

5.3.2 Pulse circuit qualification

Prior to the construction of a full-scale version of the generator, a smaller proof-of-concept version was assembled and tested.

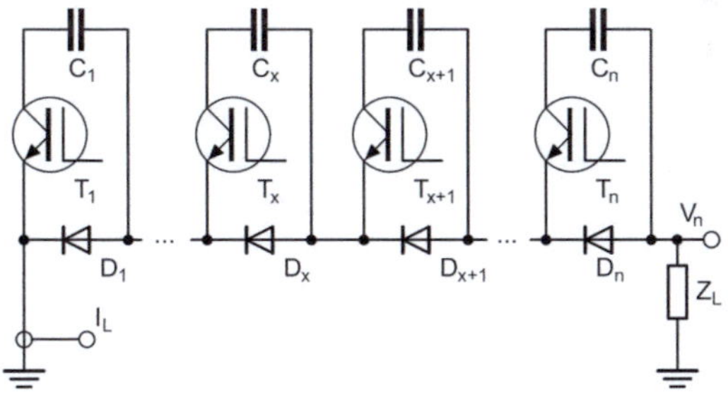

Figure 5.12: The test circuit for an arrangement of up to eight stages. The generator output was connected to a load impedance Z_L, across which the total output voltage V_n was measured. The load current I_L was measured in the grounding point of the generator.

The test setup allows for arranging up to eight stages according to the circuit diagram in Fig. 5.12. In the first step, a load impedance Z_L was connected to the output. The output voltage V_n was measured across the output impedance and the load current was measured at the grounding point of the generator (for the measurement equipment, refer to Sec. 2.5).

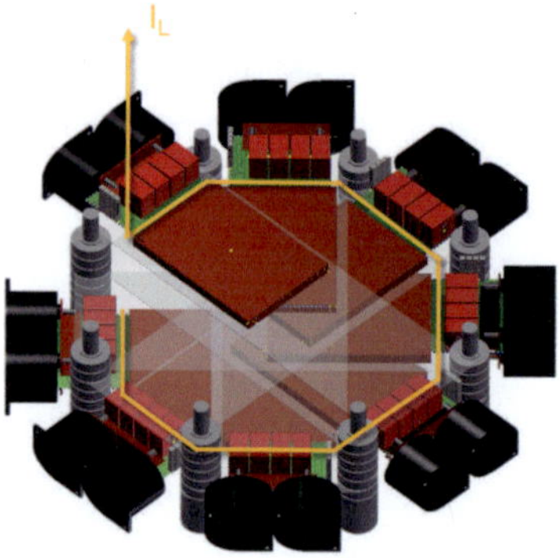

(a) CAD drawing of an eight-stage arrangement.

(b) Photo of the corresponding experimental setup.

Figure 5.13: Photo and CAD drawing of the eight-stage generator arranged in a helical pattern.

The CAD model of the arrangement together with a photo of the actual setup is shown in Fig. 5.13. As can be seen in Fig. 5.13a, the stages are arranged in a helical pattern. The load current follows the winding of the stages upwards. As resistors, electrolytic water resistors were used, connected to the topmost stage (generator output) and the bottommost stage on ground potential.
For estimating the inherent rise time of the generator (switches and jitter), the arrangement was operated at full charging voltage 1 kV in nearly open circuit conditions. The residual load impedance resulted from the connected high-voltage divider (C_L = 64 pF [47]) and a high-value resistor (R_L = 340 kΩ). The resulting measurement is shown in Fig. 5.14. All stages were triggered simultaneously, with a resulting output voltage of 8 kV. Due to the capacitive component of the load, strong oscillations are visible after the generator erects. The rise time of the output voltage can be estimated to 28 ns without load.

Following the discussion in Sec. 2.5.1, the limited rise time of the high-voltage divider will overestimate the rise time. Nevertheless, this value will be used to serve as worst-case estimation.

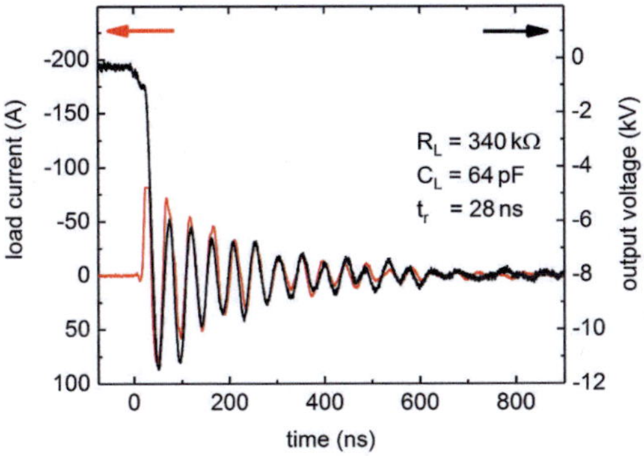

Figure 5.14: Load current and output voltage of the generator operated on a high-value ohmic load in near open circuit conditions.

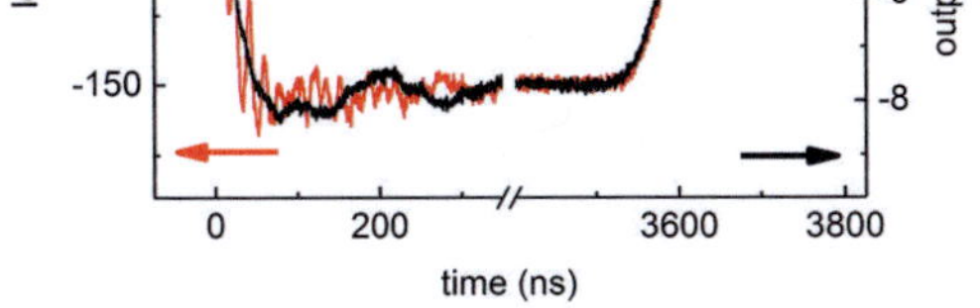

Figure 5.15: Output voltage and load current of a fast rising output pulse of approximately 8 kV, 150 A connected to ohmic load.

Fig. 5.15 displays a measurement with around 8 kV output voltage across a 50 Ω load. Since load current and output voltage coincide, the assumption of an ohmic load is validated. The majority of the remaining circuit inductance, hence, is in located in the generator arrangement instead of being part of the load resistor. For 150 A pulse current, the current and voltage rise time equals 40 ns. The falling edge being determined by the charge carrier recombination (see Sec. 3.3.4) is significantly slower than the rising edge. The asymmetry of the pulse has no influence of the generator performance in the investigated application.

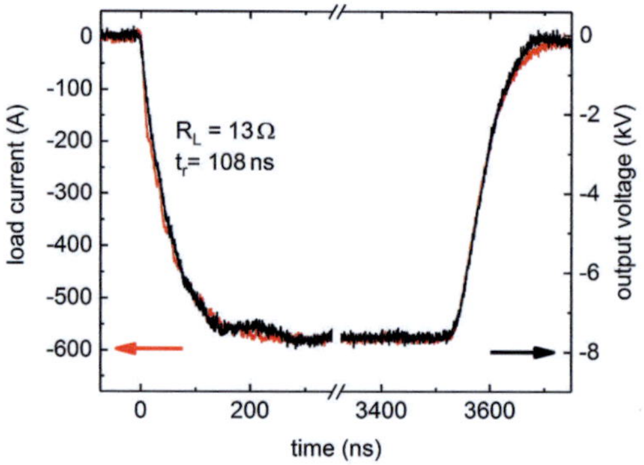

Figure 5.16: Output voltage and load current at maximum pulse parameters of around 8 kV and 600 A. By increase of the RL time constant, a low-value ohmic load slows down the generator rise time.

Running the generator at maximum pulse parameters of approximately 8 kV and 600 A when connected to an ohmic load of 13 Ω (see Fig. 5.16), the rise time increases to 106 ns due to an increase in the RL time constant. In agreement with the measurements presented in Sec. 3.3.4, the pulse symmetry increases with increasing pulse current. From these measurements, the circuit inductance can be estimated. Using Eq. 4.7, the rise time contribution caused by the RL component can be derived under the assumption of the inherent generator rise time being 28 ns. The results of both graphs Fig. 5.15 and Fig. 5.16 are summarized in table 5.3. Both measurements yield identical results within the resolution of the measurement of 1.4 μH.

Table 5.3: Calculated stray inductance values.

τ_V	Load resistor (peak current)	τ_I	Calc. circuit inductance
28 ns	50 Ω (150 A)	40 ns	1400 nH
28 ns	13 Ω (600 A)	106 ns	1350 nH

5.3.3 Arbitrary output waveform generation

The measurements presented in the preceding section were performed with eight stages switched on simultaneously. By varying the amount of active stages, however, it is possible to control the output voltage step-wise arbitrarily. The simplest case is to create a flat pulse with varying pulse amplitude. Fig. 5.17 demonstrates how the amplitude of the output waveform can be chosen in steps of the charging voltage (1 kV) from one to eight active stages.

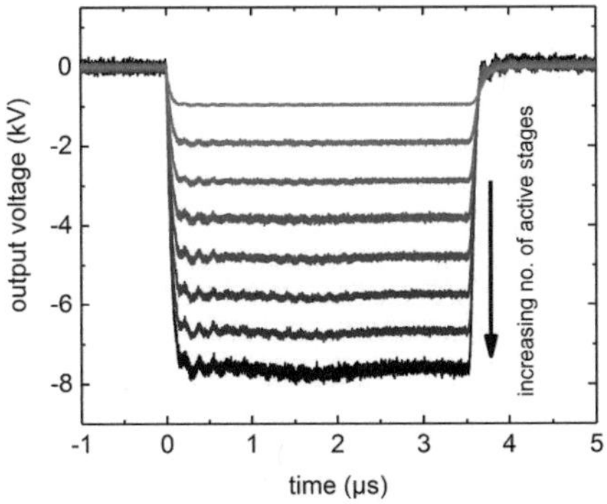

Figure 5.17: A 3.5 µs pulse at 1 kV stage voltage. By varying the number of active stages, the output voltage can be chosen step-wise arbitrarily.

Apart from flat pulses, the control of the GESA device may necessitate a more sophisticated shape of the driving voltage. The pre-programmed switching sequences each stage can perform allow for changing the output during the pulse. An example of a true step-wise arbitrary waveform is depicted in Fig. 5.18. At maximum pulse parameters of approximately 600 A and a maximum output voltage of around 8 kV, a pulse resembling the IHM institute logo can be executed. Due to the long pulse length and high current, the capacitor voltage droop is visible in the course of the pulse.

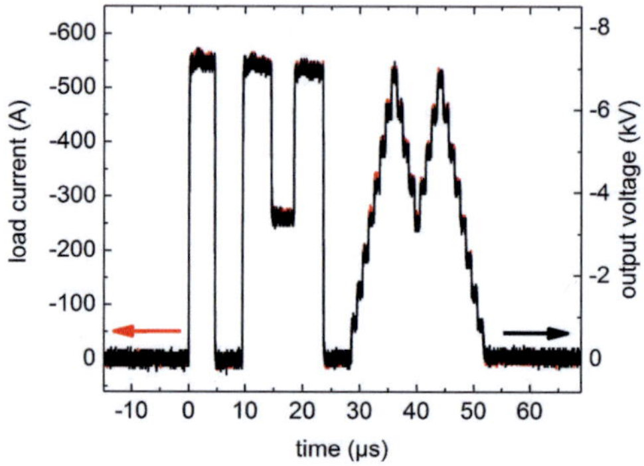

Figure 5.18: Arbitrary output waveform as delivered by the generator connected to an ohmic load, resembling the IHM institute logo.

5.3.4 Overcurrent protection

The dynamic behavior of the emitting plasma front in the GESA device render it a highly dynamic load. Especially a complete short of the cathode-anode gap may occur during experiments. For this reason, the short-circuit capability of the generator had to be tested.

Two different types of faults have been investigated in this work: a turn-on fault and a fault-under-load. The first is caused by a complete short of the load even before the full output voltage is applied. This case may happen in case of a complete flash over of the device. The second is a more likely scenario: a pulse is executed with nominal parameters when the load impedance suddenly drops strongly. Once a short circuit occurs, the load current I_L is only limited by the remaining circuit inductance L_L. It will rise linearly depending on the generator output voltage V_n according to Eq. 5.1.

$$\frac{dI_L}{dt} = \frac{V_n}{L_L} \tag{5.1}$$

Using the result of the inductance measurement of the previous section (1.4 µH/8 stages), the inductance of the final generator (149 stages) including the GESA device and wiring can be estimated to around 30 µH (see Sec. 6.1). With a nominal generator output voltage of 120 kV, the expected current rise rate will be in the order of 4 kA/µs. To model these conditions, the load impedance as introduced before was expanded by an inductance and a spark-gap in parallel (see Fig. 5.19). Upon the ignition of the spark-gap, the load changes its behavior from resistive to inductive. For the following experiments, the circuit as presented in Fig. 5.13 was operated with seven stages. For a total output voltage of 7 kV, the required circuit inductance to achieve a rise rate of 4 kA/µs is around 1.75 µH. Subtracting the internal inductance of the seven-stage arrangement of 1.23 µH, the load inductance L_L was chosen to 520 nH.

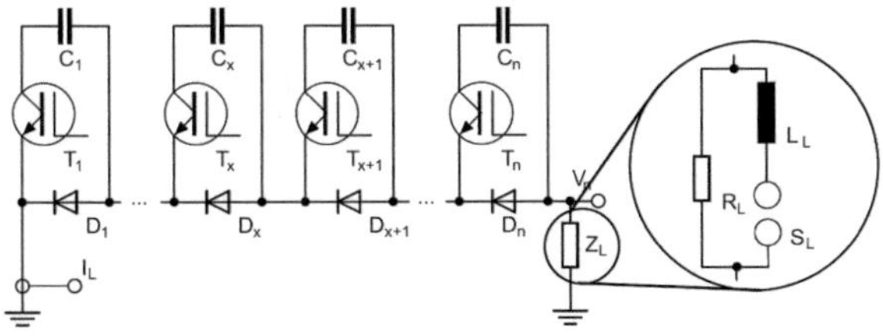

Figure 5.19: Expanded circuit diagram of the load impedance. For testing the behavior on dynamic loads, an additional inductance can be connected in parallel to the load during the pulse by means of a spark gap.

For the turn-on fault, the spark-gap was shorted and the generator fully charged to 1 kV stage voltage. All stages were triggered simultaneously. The resulting current measurement is depicted in Fig. 5.20a.

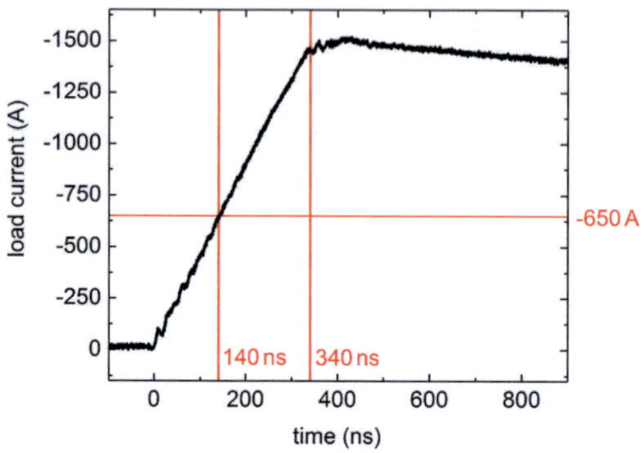

(a) Current measurement of the generator turning on at full charging voltage onto a low-inductive short.

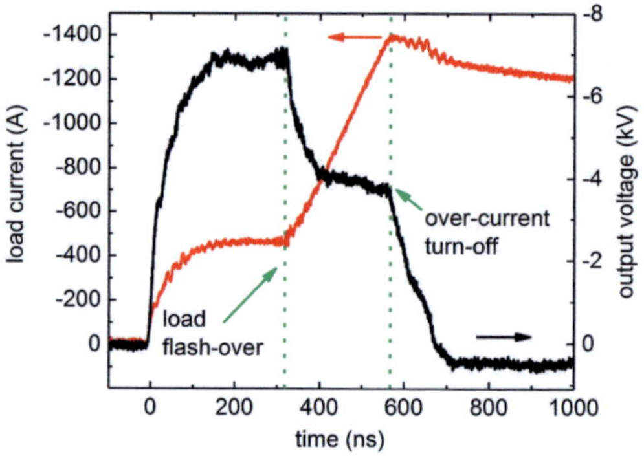

(b) Dynamic load testing of the generator: the load is shorted around 300 ns after pulse start.

Figure 5.20: Current and voltage traces during the successful clearance of overcurrent faults in different scenarios.

The current rises linearly with a slope slightly higher than 4 kA/µs. After 140 ns, the threshold current of 650 A is reached. It takes around 200 ns for the generator to register the overcurrent condition and turn off. During this time, the load current crests at 1.5 kA. Considering the maximum pulsed current the switching elements can sustain (1.9 kA), the safety margin is sufficient to

protect the devices. The 200 ns time delay between reaching the overcurrent threshold and generator turn-off is caused by the current measurement and the logic circuitry (approximately 100 ns) and the time delay from the removal of the gate signal until the IGBTs start blocking (approximately 100 ns).
The second scenario, the fault-under-load, is depicted in Fig. 5.20b. The spark-gap was operated close to its self-breakdown voltage and ignites with a time delay. For this measurement, a flat pulse of 7 kV and 400 A was executed. After the load flashes over (i.e. the spark gap ignites), the load current again rises linearly until the generator turns off. The maximum current of 1.4 kA is still in the safe range. With these measurements it was demonstrated, that the generator is able to withstand short-circuit conditions under different fault scenarios. It has to be noted that the maximum current rise rate has to be limited to values below 4 kA/µs in order to prevent damage to the device due to the dead-time between reaching the overcurrent threshold and generator turn-off. As described in Sec. 6.2.2, a current-limiting inductance is an option to reduce the current rise rate while allowing a fast charging process of the generator's stray capacitance.

6 Generator implementation

After the successful tests with a small-scale assembly presented in the preceding section, the full-scale version of the generator was developed and constructed. The main circuit parameters resulting from the increased physical dimensions as compared to the smaller version can be estimated by 3D electromagnetic simulations with a simplified model and used for worst-case estimations. Measurements on the full-scale version validate the capability of 120 kV output voltage while performing active droop compensation. The strict rise time requirement of 10^{12} V/s is thereby exceeded by 20 %. At nominal current of 250 A, a maximum pulse length of 100 μs can be achieved. Selected results of this work presented in the following chapter have already been published as [23].

6.1 Electrical properties

The high number of individual stages, each comprising a complex circuitry, quickly escalates the required effort for a detailed simulation. In order to estimate the generator's behavior before construction, important values such as generator inductance and stray capacitance to ground, have to be calculated. Whereas both of these values are distributed elements with a frequency dependency, simple worst-case estimations have been used in the design of the generator. Distributed elements can be considered "lumped" if signal transit time over the extend of the element is significantly faster than the time of the energy transfer with the element [15]. With the large employed stage capacitances, the energy transfer time is in the microsecond range, thus exceeding the signal transit time by at least one order of magnitude. Therefore, the total circuit inductance and stray capacitance can be considered as a lumped element as far as the main pulse is concerned. Fast oscillations across distributed elements, excited by the fast transients of the switching elements cannot be identified using this method.

6.1.1 Stage arrangement

In the final assembly, the stages are arranged in a similar pattern as investigated in the previous section. Fig. 6.1a shows a CAD model of around 40 stages.

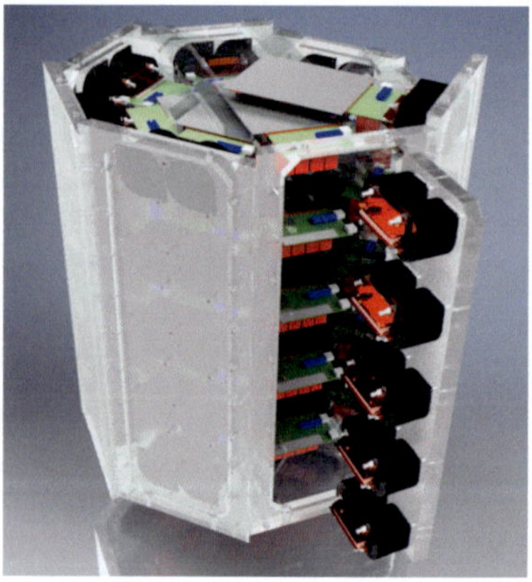

(a) The stages are arranged in a helical pattern comparable to the test assembly (Fig. 5.13). The 100 µF capacitors are mounted on doors to allow for an easy access.

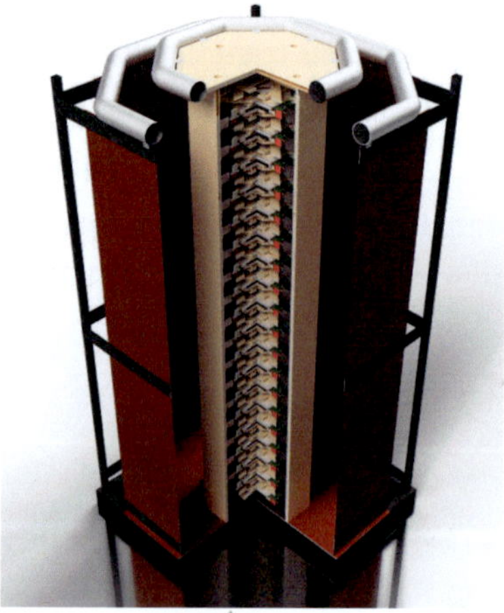

(b) With the stages enclosed in the inner tower and the return conductor, consisting of copper plates, arranged around it the generator is designed to be a quasi-coaxial arrangement. The output is connected between the two top field grading rings.

Figure 6.1: Illustration of the stage arrangement within the octagonal tower (a) and generator including the return conductor (b).

They are arranged in a helical pattern as close as possible together. The main spacial constraint being the pulse capacitors, they are mounted facing outwards. The walls of the structure are sectioned in doors with the capacitors mounted on them. Since the capacitors are connected to the PCB via contact springs only, they can compensate a certain amount of misalignment and the doors can be opened and closed (capacitors connected and disconnected) without the need of tools. For easy maintenance, the boards are only slid into the copper shielding enclosures and can be exchanged quickly. Additionally, since each board features a 4 µF capacitance mounted directly on the board, the generator can be operated without the main pulse capacitors during commissioning. Several of the units as depicted in Fig. 6.1a are arranged into a tower as can be seen in Fig. 6.1b. The generator output is connected to the smaller field grading ring at the top. For reduced inductance of the circuit, the ground conductor is formed by copper plates all around the generator. Connecting the load between the two field grading rings, the generator is designed to be quasi-coaxial.

6.1.2 Stray inductance estimation

Arranging the stages in a helical pattern to achieve a low-inductive current path seems counter-intuitive due to the increased inductance resulting from an magnetic field enhancement caused by the parallel current paths. However, the important factor determining the magnetic field enhancement is the field overlap from on turn to the next.

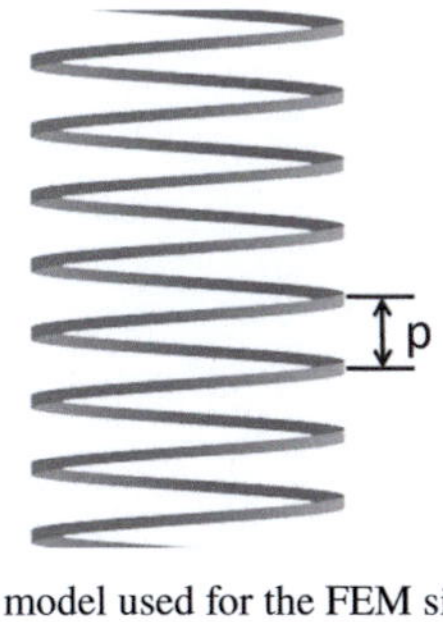

(a) The model used for the FEM simulations: a helical current path with variable pitch p.

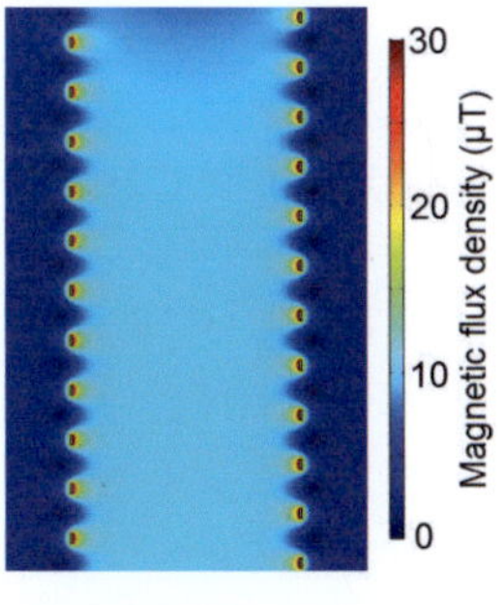

(b) Using COMSOL (magnetic field module), the magnetic flux density for a DC current of 1 A and a pitch of 50 mm shows a significant overlap from one turn to the next.

Figure 6.2: Model and resulting magnetic flux density for estimating the inductance of a helical current path with pitch p.

In order to determine the resulting inductance, static magnetic field simulations were performed (COMSOL, AC-DC module) on a helical current path with variable pitch p and otherwise identical geometrical dimensions. The used model is shown in Fig. 6.2a. The physical dimensions are equal to the presented design. For a small pitch of 50 mm and a DC current of 1 A, the resulting magnetic flux density is depicted in Fig. 6.2b. The small pitch causes a significant magnetic field overlap. From the resulting total magnetic energy, the circuit inductance was calculated.
Fig. 6.3 summarizes the results of the simulations. The total inductance value was divided by the number of necessary turns in the generator (18) as figure-of-merit but, therefore, is only a mean value. As can be seen, the inductance increases strongly for small pitch values. Above a pitch of 200 mm, the change of the inductance with increasing pitch is far less pronounced. For high pitch values, the physical dimensions of the generator become impractical. The red star indicates the value as chosen in the design as trade-off between circuit inductance and generator size. The value of 1.4 μH per eight stages is in good agreement with the measurements presented in the proof-of-concept assembly.

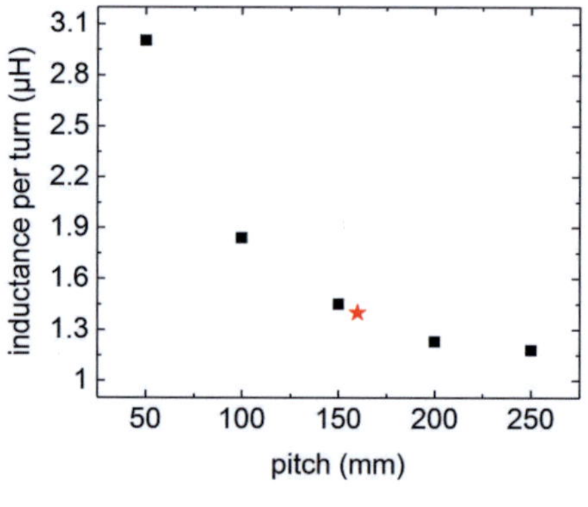

Figure 6.3: Simulated values of the generator inductance per turn (149 stages equal 18 turns) in dependence of the stage pitch. The red star indicates the chosen value of 160 mm as best trade-off between generator size and inductance.

Since DC simulations neglect the effect of eddy-currents induced in metal structures present in the generator (reducing the inductance), the actual value of the generator inductance can be much lower. Additionally, the simulations disregard the influence of the ground conductor on the arrangement. However, as worst-case estimation this value can be used for generator design.

6.1.3 Simulating the parasitic capacitance

The generator's stray capacitances are distributed elements over the whole extend of the generator. For each stage of the generator, a stage-to-ground capacitance C_{SG} and a stage-to-stage stray capacitance C_{SS} can be defined according to Fig. 6.4.

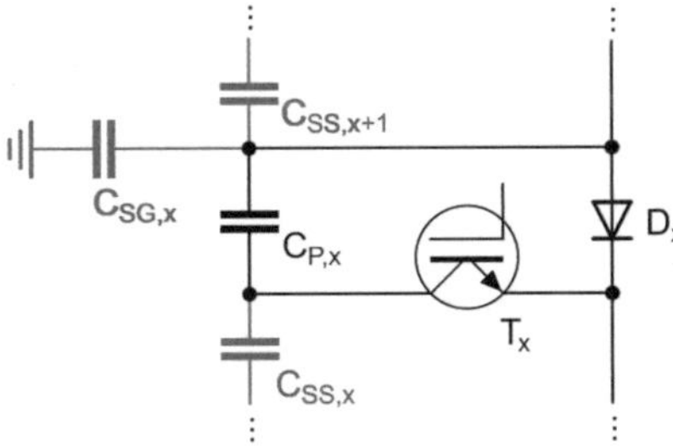

Figure 6.4: Equivalent circuit of stage, including the stray capacitance towards ground $C_{SG,x}$ and a stage-to-stage capacitance $C_{SS,x}$.

In principle, a high distance between the stages and ground decreases the stray capacitance but simultaneously increases the circuit inductance. To avoid the time-consuming simulations with a detailed generator model, a simplified model was used. Therefore, the CAD model (see Fig. 6.5a) was simplified by removing the black support structure of the copper return conductors. Instead of modeling every PCB and capacitor inside the octagonal tower, it was simulated empty. A linear potential distribution inside of the tower was enforced by filling it with a high-permittivity material (its electric field energy contribution was subtracted afterwards). Therefore, the complete inner surface of the tower contributes to the stray capacitance, overestimating it. As boundary conditions, the top cover of the generator tower was simulated as electrode at V_n = 120 kV, the base plate of the generator and the return conductor were grounded. Instead of calculating each stray capacitance value (differing from one stage to the next depending on the position), the total electric field energy

$E_{E,total}$ was calculated. Using Eq. 6.1, the equivalent generator-to-ground capacitance C_{GG} can be calculated.

$$C_{GG} = C_{SG,total} = \frac{2 \cdot E_{E,total}}{V_n^2} \tag{6.1}$$

In Fig. 6.5b, the electric field distribution inside the generator is depicted. A total electric field energy of 800 mJ yields an effective generator capacitance of around 100 pF.

(a) CAD model of the generator including the return conductor.

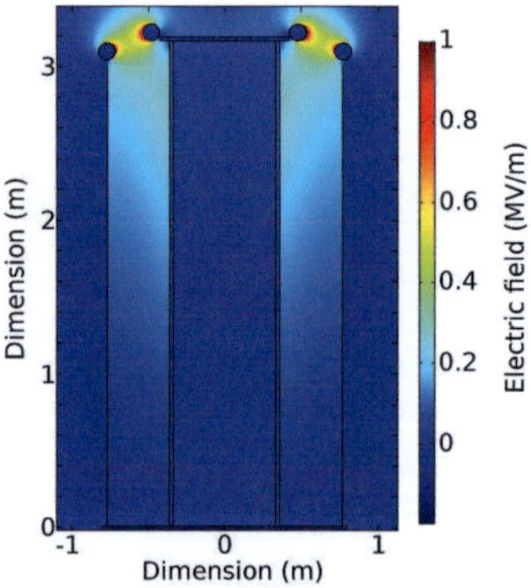

(b) Electric field distribution for the nominal output voltage of 120 kV and a constant voltage gradient within the generator tower.

Figure 6.5: CAD model and electric field distribution used for deriving the equivalent stray capacitance to ground of the generator.

The stage-to-stage capacitance can be estimated assuming a parallel-plate capacitor. The overlapping area of one stage to the next is in the order of 280 cm^2 with an air gap of 1 mm. The small potential difference of 1 kV yields for all 149 stages a total stored energy of 18 mJ. Due to the small energy stored in $C_{SS,total}$ as compared to $C_{SG,total}$, the stage-to-stage capacitance is neglected in the following calculations.

6.1.4 Generator connected to the GESA device

Before the final construction of the generator, the design needs to be validated using all worst-case calculations for each component as presented earlier. The whole circuit comprised of the generator, its connection to the GESA device and the GESA device itself. By combining all parasitic capacitances and parasitic inductances in two lumped elements that are to be driven by an ideal voltage source with limited rise time, a lower boundary for the voltage rise time across the GESA cathode can be derived. For connecting the generator output to the GESA device, a coaxial line of 2 m length is to be used. With an inner diameter of 5 mm (150 kV DC insulated line) and an outer diameter of 8 cm, it has a parasitic inductance of 1 µH and a parasitic capacitance of 40 pF. Table 6.1 summarizes all parasitic properties in the arrangement.

Table 6.1: Summary of the parasitic properties as used for the calculation of the cathode voltage rise time.

Property	Value
Generator rise time	30 ns
Generator capacitance	100 pF
Generator inductance	25 µH
Coaxial connector capacitance	40 pF
Coaxial connector inductance	1 µH
GESA capacitance	150 pF
GESA inductance	4 µH

The inset in Fig. 6.6 shows the equivalent circuit of the arrangement: all capacitive contributions are combined in one element C_C (190 pF) whereas all inductive contributions are combined in L with 30 µH. The dashed trace in Fig. 6.6 represents the ideal generator with a rise time of 30 ns up to 120 kV. This value equals the value obtained in the small-scale assembly (see Sec. 5.3.2). The black trace indicates the voltage across the capacitive element, rising in significantly less than 100 ns (91 ns, gradient: 10^{12} V/s) to 100 kV (red dashed lines indicating the 10 % and 100 % thresholds).

With a current of less than 400 A necessary to charge all parasitic capacitances, all design values are met.

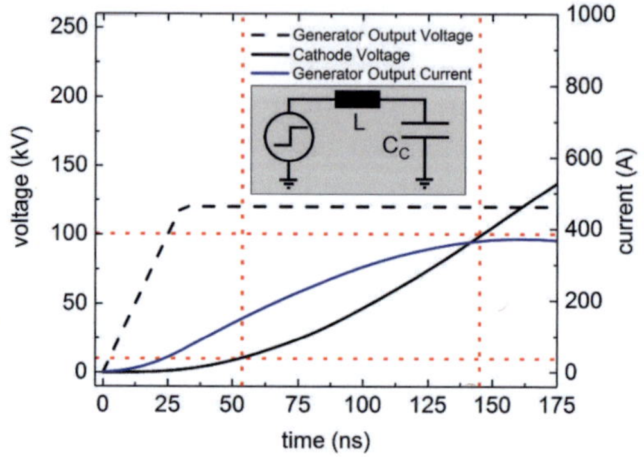

Figure 6.6: Simulated cathode voltage according to the equivalent circuit as shown in the inset. Combining the worst-case estimations for the generator and GESA capacitance, generator and wiring inductance and limited generator rise time, the voltage rise rate requirement of 10^{12} V/s can be met.

It has to be noted, that the voltage overshoot as expected in a weakly damped RLC circuit does not cause problems within the device as the device draws a significant current as soon as the cathode plasma ignites (above 100 kV). Whereas the simulations verify that the design goals can be met even under worst-case estimations, the presented design was implement. The actual circuit parameters, however, have to be determined by experiments once the complete system is set up.

6.2 Full-scale validation

In order to benchmark the generator's performance, it was operated in a test-bed for the following experiments. For the implementation of the device into the GESA experiment, a time-consuming rearrangement of the experiment as well as modifications to its control system have to be performed. This task will be completed by the GESA experimentalists in the near future. The photos presented in Fig. 6.7 show the generator after construction without (a) and with (b) return conductor.

(a) Generator, power supplies and controls (left side) after construction.

(b) Generator with the return conductor mounted.

Figure 6.7: The photos show the final implementation of the generator in its test-bed. Some of the following experiments were performed without the 100 μF capacitors, therefore the generator was operated without the doors.

6.2.1 Hardware discharge switch

For a safe operation of the generator, all capacitors need to be shorted after experiments. Whereas each stage features a discharge switch (see Sec. 4.2.4), it is not considered safe since it is controlled by the logic circuitry. In addition to the contact springs shorting the capacitors when the doors are opened (see Sec. 4.1.2), a second mechanical switch is required.

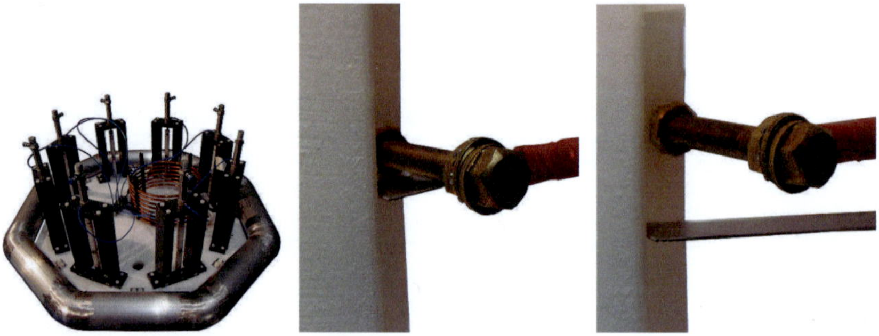

(a) The pneumatic pistons located at the top of the generator operate eight individual shorting switches by means of rods.

(b) When turned off, the capacitor lead connected to the spring steel lever and the lead connected to the brass bolt are shorted.

(c) During operation, the pistons lift the rod, enabling a charging of the capacitor.

Figure 6.8: Photos showing the implementation of 149 individual capacitor discharge switches operated simultaneously by pneumatic pistons. For each stage, the pulse capacitors are connected to a spring steel cantilever and brass bolt (b) and (c).

Due to the high number of stages, an automated mechanism was installed. Operated by pressurized air, eight pistons operate eight individual insulating rods running along the generator. The rods are pulled down by means of a metal plate at the bottom of the generator and can be lifted by 3 cm by pressurizing the pistons. For each stage, one pole of the capacitor is connected to a metal contact mounted on the moving rod, whereas the other pole is connected to a fixed spring steel cantilever. Without air pressure, the contact rests on the cantilever and the capacitor is shorted. During operation, the pistons lift the rods and break the electrical contact. The valve controlling the air pressure is operated by the experiment's safety circuit. In case of an emergency shutdown, the power supplies are turned off. The lack of control power causes a closing of the local discharge switches on each stage, so that each capacitor is safely discharged via the discharge resistor before the hardware discharge switch causes a hard short of the capacitors. The system operates in less than 2 s.

6.2.2 Determination of pulse parameters

In order to evaluate the generator after construction, its pulse parameters have to be determined. Operating the generator without the 100 µF pulse capacitors and pulsing only one stage enables a simple method to determine the generator stray inductance: at full charging voltage, one stage discharges its pulse capacitor into a short-circuit. The equivalent circuit of the generator can be seen in Fig. 6.9. The equivalent capacitance C_{eq} is derived from the series connection of all active stages – in this experiment it equals the total capacitance of one stage (4 µF) – being charged to the sum of the charging voltage of all active stages V_n (1 kV). Since the load current by-passes all inactive stages via the free-wheeling diode, the series connection of all free-wheeling diodes is represented by an equivalent diode D_{eq} and its corresponding voltage drop V_D. All losses are accounted for by the incorporation of a parasitic resistance R_P. Additionally, the circuit has a parasitic inductance L_P.

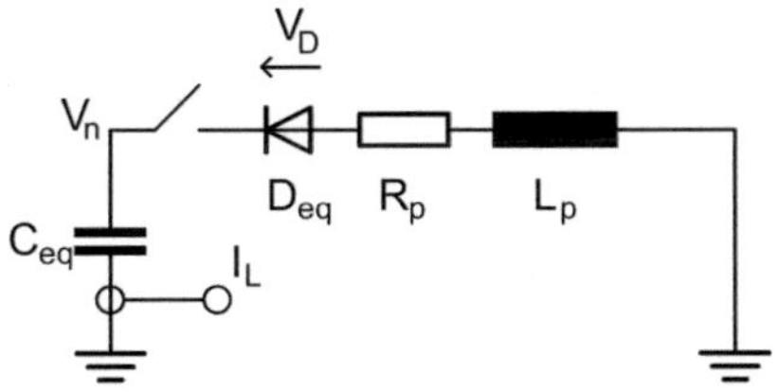

Figure 6.9: The generator's equivalent circuit in case of a load short. The equivalent capacitance and equivalent diode parameters are determined by the number of active stages, parasitic inductance, and parasitic resistance are parameters of the circuit.

Fig. 6.10 shows the capacitor voltage of one stage (red) and the resulting current waveform (black). As the arrangement is a basic RLC discharge, the current trace can be fitted to extract the circuit parameters. The orange dashed line represents the current fit using Eq. 6.2.

$$I_L = \frac{V_0}{\omega \cdot L_P} \cdot \exp(-\delta \cdot t) \cdot \sin(\omega \cdot t)$$

$$\text{with: } \delta = \frac{R_P}{2L_P}, \; \omega = \sqrt{\frac{1}{L_P C_{eq}} - \delta^2} \text{ and } V_0 = V_n - V_D \tag{6.2}$$

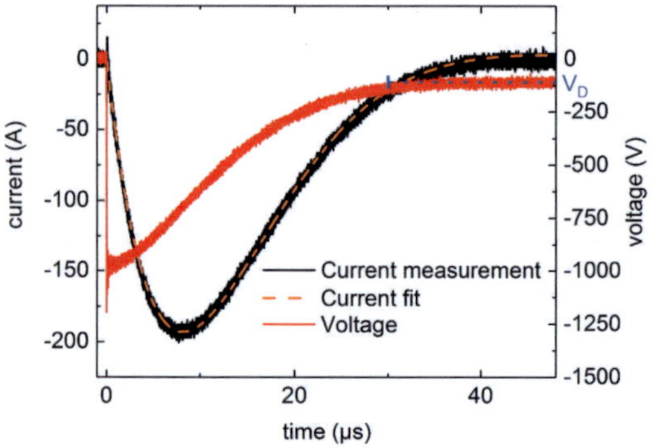

Figure 6.10: Output voltage (red) and current (black) when pulsing only one stage in the generator onto a short circuit. From the current fit (yellow dashed), the circuit parameters can be determined.

As can be seen in the graph, the pulse capacitor does not discharge completely due to the finite diode forward voltage V_D. The value can be extracted to 115 V from the measurement. The resistive component of the circuit as used in the fit has a value of 3.1 Ω, the inductive component amounts to 14.8 µH. The rather high value of the resistive component can be explained by the differential resistance of the free-wheeling diodes. Each diode has a differential resistance of 100 mΩ (10 A, 25 °C [88]). Using six diodes in parallel, the differential resistance of each stage in by-pass mode equals approximately 16.7 mΩ. Resulting from the series-connection of 148 inactive stages in this measurement, the ohmic contribution of all diodes amounts to approximately 2.5 Ω. The residual 600 mΩ of resistance, if distributed equally across all stages, amount to 4 mΩ per stage. Compared to the worst-case estimation in Sec. 6.1.2, the inductance of the final generator was found to be only 60 % of the simulated value. A significantly lower value was expected due to the simplification of a static simulation and the resulting neglect of effects such as eddy-currents reducing the inductance. This result has an important impact on the generator design, as the much lower stray inductance allows for current rise rates exceeding the capabilities of the implemented overcurrent protection. To limit the current rise rate below 4 kA/µs (refer to Sec. 4.2.3) in case of a load flash-over, an additional current-limiting inductance of 16 µH was connected between the last stage of the generator and top field-grading ring.

6.2.3 Measurement qualification

For a complete understanding of the measurements, all before mentioned parameters have to be considered. To this end, the equivalent circuit as presented in Fig. 6.9 has to be extended according to Fig. 6.11. Load voltage V_L and load current I_L are measured at the load resistor R_L.

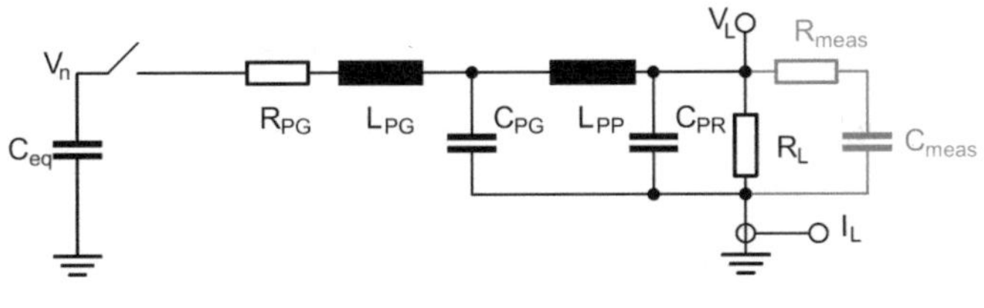

Figure 6.11: The equivalent circuit of the Marx generator attached to an ohmic load. The effects arising from the connected capacitive-resistive high-voltage divider are incorporated (gray components).

The equivalent capacitance C_{eq} = 33 nF is calculated as series connection of all 120 active stages, charged to voltage V_n = 120 kV as sum of all charging voltages. The switch is modeled to close within 30 ns. At full output voltage (i.e. 120 kV, 120 stages active), only 29 stages will be in by-pass mode with their diodes active. Due to the high output voltage, their forward voltage drop can be neglected. The resistive component of the generator will be then determined predominantly by the differential resistance of the switching elements. For one stage, the differential resistance equals 2 mΩ (100 A per device, 25 °C [67]), in combination with the calculated 4 mΩ contact resistance, the active generator has a parasitic resistance R_{PG} of approximately 1 Ω. The parasitic generator inductance L_{PG} amounts to 14.5 μH and the protection inductance L_{PP} to 16 μH (see Sec. 6.2.2). As L_{PP} separates the stages from the top field-grading ring, the generator's total capacitance is split into one component of all the circuitry in the tower C_{PG} and one component C_{PR} (i.e. the field-grading ring) after the protection inductance, including the coaxial connection to the load resistor (20 pF). Using the simulations shown in Fig. 6.5b, the two components can be calculated to 40 pF and 80 pF. For first experiments, the generator was connected to a load resistor R_L of 1400 Ω. In addition to generator and load, also the measurement system, a capacitive-resistive high-voltage divider (HVT 240 RCR – HILO TEST), has to be considered. Its components have been included according to the datasheet values of R_{meas} = 114 Ω and C_{meas} = 190 pF [47].

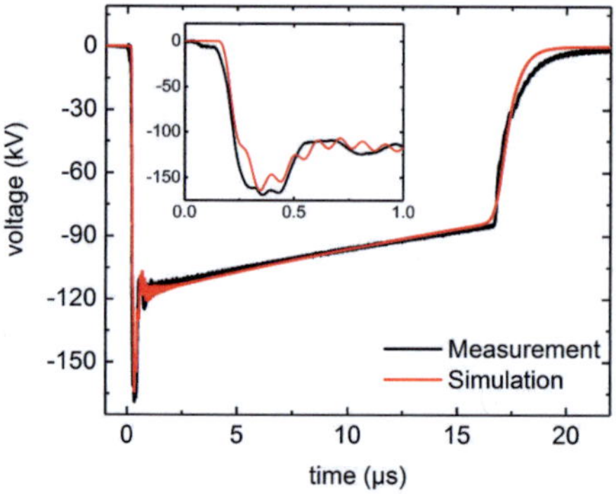

Figure 6.12: Measurement (black) and corresponding simulation (red) of a 120 kV output pulse. The inset shows a magnification of the leading edge.

A comparison of the simulated voltage across the load (SPICE, circuit Fig. 6.11) and a measurement is shown in Fig. 6.12 for a pulse duration of 16 μs. With 120 active stages charged to 1 kV charging voltage, a strong overshoot at the load is visible. A magnification of the leading edge (inset) shows a good agreement between measurement and simulation. The voltage overshoot is caused by the significant inductance in the circuit in combination with an important capacitive contribution from the measurement system. Comparing the load current and load voltage measurement as in Fig. 6.13, the absence of an important inductive component in the load is verified by the coincidence of both traces.

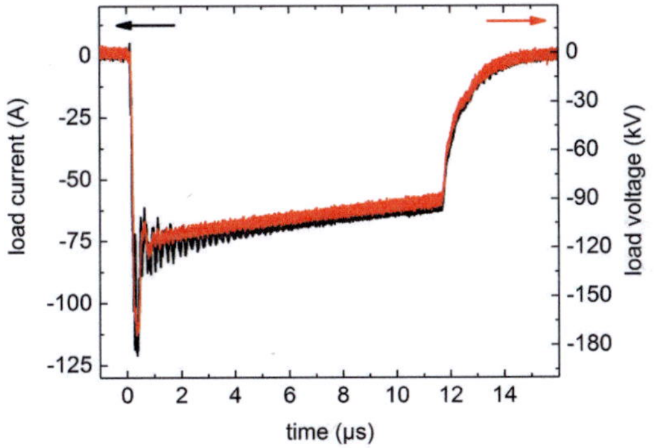

Figure 6.13: Connected to a high-value ohmic load of 1.4 kΩ, both load current (black) and load voltage (red) show a strong overshoot.

6.2.4 Meeting the voltage rise time requirement

In the final arrangement, the voltage rise time across the GESA cathode will be the main figure of merit of the generator. As presented in Sec. 2.1.3, the cathode has a parasitic capacitance of around 150 pF which is in the same range as the parasitic capacitance of the measurement system (190 pF). Since the GESA device features a build-in resistive high-voltage divider, the measurement system used for the presented experiments will not be used in the future. Therefore, the voltage rise time as measured in the presented system will be only insignificantly slower than the voltage rise time across the GESA cathode. Fig. 6.14 shows a magnification of the leading edge as presented in Fig. 6.12 together with a linear approximation of the rising edge. The rise time from 10 kV to 100 kV is 78 ns as discussed in Sec. 2.1.2. Using the 10 % to 90 % definition, the rise time to peak output voltage is 97 ns. As mentioned in Sec. 2.5.1, the limited rise time of the high-voltage divider of 48 ns has an influence on the measured rise time. Under the assumption that the influence of the divider can be taken into account according to Eq. 2.18, actual generator rise time can be calculated to 84 ns. Since both systems (generator and high-voltage divider) interact significantly, this calculation may be misleading. However, as the measured signal already shows a voltage gradient of $1.2 \cdot 10^{12}$ V/s and thereby exceeds the requirement of 10^{12} V/s by 20 %, conditions for a homogeneous plasma ignition are achieved. The strong overshoot above the nominal 120 kV drive voltage is not considered a problem, since the GESA load behavior changes from capacitive to resistive as soon as the cathode plasma ignites. All energy stored in the parasitic inductances causing the voltage overshoot will be dissipated in the device.

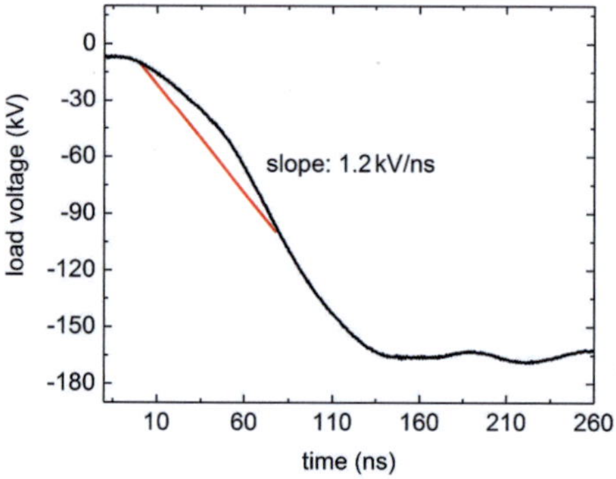

Figure 6.14: A magnification of the pulse as presented in Fig. 6.12. The voltage rise rate across the load (resistor and capacitive divider) is deduced from the rise time from 10 kV to 100 kV.

6.2.5 A 120 kV, 600 A semiconductor-based Marx generator for step-wise arbitrary output waveforms

At full charging voltage, the stage's switching program can be chosen arbitrarily to achieve a modulated output waveform. To limit the energy input into a sample, the pulse length has to be adjustable. As example, Fig. 6.15 displays pulses with 7 µs, 12 µs and 17 µs pulse length at 120 kV at 1.4 kΩ ohmic load.

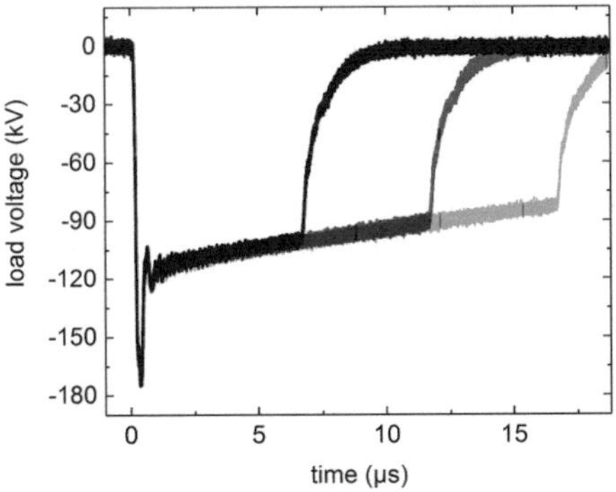

Figure 6.15: Load voltage for different pulses. The pulse length is determined by software.

By choosing the number of active stages, the total output voltage can be varied in steps of the stage charging voltage (i.e. 1 kV). Fig. 6.16 shows a comparison of 100 µs pulses with increasing active stage number (in steps of ten stages) onto 1.4 kΩ load ohmic load. As can be seen, the effective circuit capacitance decreases with increasing stage number due to the series connection of the stage capacitances.
From previous measurements it is known that the initial resistance of the GESA device is in the order of 480 Ω. However, the generator is able to turn-on onto a comparably low-impedance load. Fig. 6.17 demonstrates the generator turn-on onto 200 Ω ohmic load. The resulting pulse current of 600 A is the design limit of the generator. For the measurement, the pulse is started with 120 active stages and shows significant voltage droop. After 2.5 µs, 20 spare stages are activated to demonstrate the arbitrary waveform generation, causing the generator output voltage to increase by approximately 20 kV.

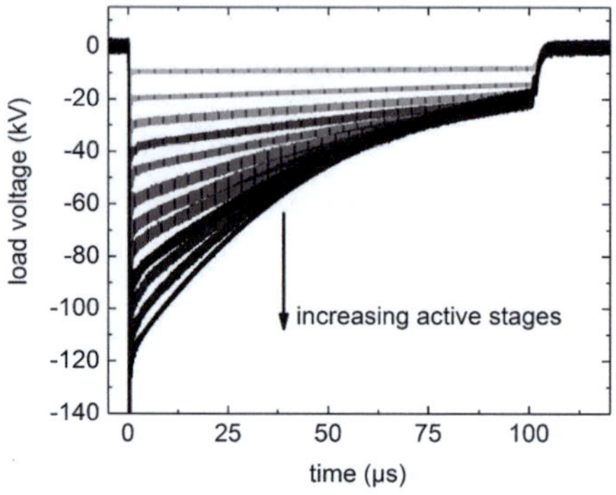

Figure 6.16: The measurement shows load voltage traces for an increasing number of active stages.

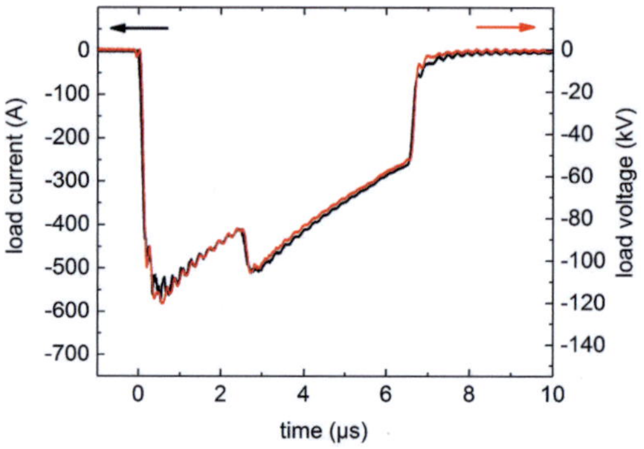

Figure 6.17: The graph shows load voltage (red) and load current when executing a 120 kV pulse onto 200 Ω load. The maximum specified current of 600 A can be achieved. After 2.5 µs, 20 spare stages are activated.

In the final stage of the project, the generator was completed by mounting the 200 µF stage capacitors. At the nominal pulse current of the GESA device of 250 A a pulse with a length of 100 µs can be achieved with minimal voltage droop. The measurement presented in Fig. 6.18 shows current and voltage traces of the generator connected to a 480 Ω ohmic load. The inset in the graph shows a magnification of the leading edge, validating the implicit assumption that the generator performance remains unchanged by the increased stage capacitances.

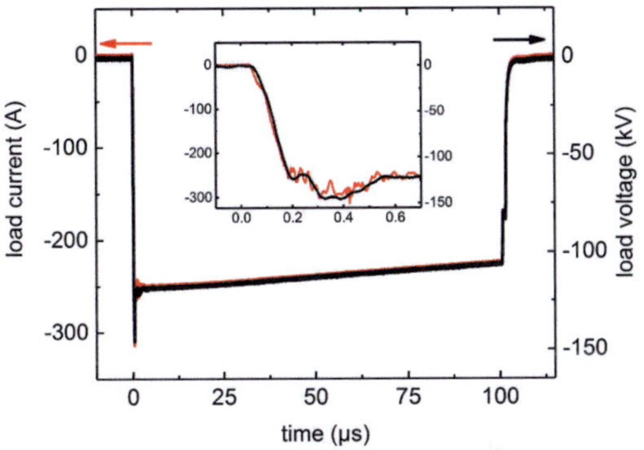

Figure 6.18: With full stage capacitances of 200 µF, a pulse length of 100 µs can be reached at nominal current of 250 A. The inset shows a magnification of the leading edge.

6.2.6 Active droop compensation

The demonstrated capability of the generator to deliver arbitrary output waveforms can be used to compensate the capacitor voltage droop by the subsequent activation of spare stages, allowing for an efficient reduction of pulse capacitor size (see Sec. 2.3.3). As an example, a comparison between a pulse without (black trace) and with (red trace) active droop compensation can be seen in Fig. 6.19. The pulse starts with 120 active stages onto a 1.4 kΩ ohmic load in both cases (4 µF stage capacitance). The series connection of 120 active stages yields the equivalent generator capacitance C_{equ} to 33 nF. After 5 µs, when the droop compensation is activated, the load current I_L has decayed to around 70 A. Allowing an output voltage droop ΔV_n of 1 kV, the time delay between the activation of a spare stage Δt_S can be calculated according to Eq. 6.3. The equation approximates the current decay to be linear. With each additional stage, the equivalent capacitance decreases.

$$\Delta t_S = \frac{C_{eq} \cdot \Delta V_n}{I_L} \tag{6.3}$$

For the measurement as presented in Fig. 6.19, hence, a time delay of around 500 ns between the activation of spare stages was chosen. Employing this technique the first standard deviation of the voltage in the range between 5 µs to 15 µs could be decreased from 5.2 kV to 0.7 kV (measurement noise: 0.15 kV).

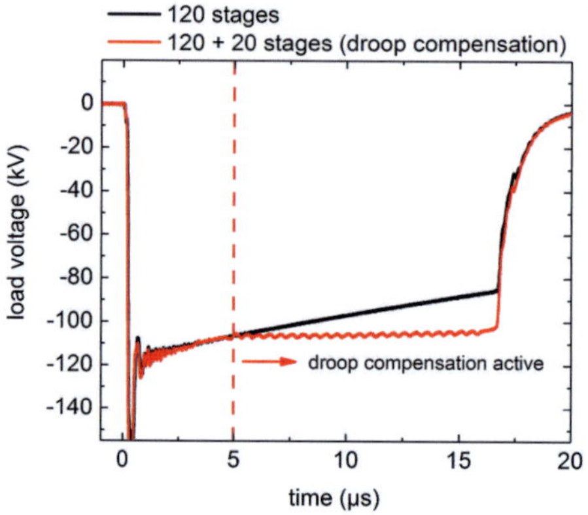

Figure 6.19: Measurement of a 120 kV pulse without (black) and with droop compensation (red).

7 Conclusion and outlook

In this work, the design of a high-voltage pulsed power modulator for fast-rising arbitrary waveforms was presented. Besides the developed "gate-boosting" technology allowing for a tremendous increase in switching speed for commercial power semiconductors, the work presents a new design to operate six devices in parallel with a low-inductive capacitive energy storage. In combination with a good synchronization and an overall low-inductive layout, the generator is significantly faster than comparable generators previously presented in literature for the given output voltage and output current. The efficient control of many stages is accomplished via a new optical bus system, allowing the transmission of data as well as the synchronization signal. In combination with the design of locally controlled switches, the constraints both on the control circuitry and the manufacturing effort have been drastically reduced. Additionally, the direct switch control allows for a fast overcurrent protection scheme. The unique EMI shielding did prove to protect the sensitive circuitry efficiently. First measurements on the full-scale implementation showed the excellent agreement with the predicted pulse parameters and the exceedance of the required rise time by at least 20 % while demonstrating the full output voltage of 120 kV and maximum load current of 600 A. At nominal pulse current of 250 A, a pulse length of 100 µs has been achieved. An active droop compensation by subsequent activation of spare stages did allow for enhancing the pulse flat top quality.

Although the efficiency of the overcurrent protection scheme was demonstrated in the small-scale assembly, the generator's behavior with full capacitance under a load-short event has to be evaluated. For future implementation into the GESA experiment, a second output of the generator will be necessary to directly control the grid voltage in the triode structure. As a result from the comparably small current drawn from the grid connection, the voltage distribution should not be altered significantly by unequal discharging of the pulse capacitors. Nevertheless, proper functioning of this concept has to be demonstrated. For the subsequent integration of the generator into the GESA

experiment, the software and hardware side of the control unit is already prepared. First measurements on the GESA device can therefore be expected in the near future.
Due to the fast rise time of the generator in combination with a high stage capacitance, the applicability of the design in other pulsed power fields of research (such as accelerator research or treatment of biological material) should be investigated. However, pulsed power generators are typically designed specifically for each application, so an adaptation of the design in details will be necessary. Especially an upgrade of the generator to higher pulse repetition rates – currently limited by the employed high-voltage relays – is a promising approach to broaden the technology's field of application.

Bibliography

[1] A. Weisenburger, G. Rizzi, A. Scrivani, G. Mueller, and J. R. Nicholls, "Pulsed electron beam treatment of MCrAlY bondcoats for EB-PVD TBC systems part 1 of 2: Coating production", *Surface and Coatings Technology*, vol. 202, no. 4-7, pp. 704–708, 2007.

[2] C. Fazio, V. P. Sobolev, A. Aerts, S. Gavrilov, K. Lambrinou, P. Schuurmans, A. Gessi, P. Agostini, A. Ciampichetti, L. Martinelli, S. Gosse, F. Balbaud-Celerier, J. L. Courouau, A. Terlain, N. Li, H. Glasbrenner, J. Neuhausen, S. Heinitz, L. Zanini, Y. Dai, M. Jolkkonen, Y. Kurata, T. Obara, N. Thiolliere, F. J. Martin-Munoz, A. Heinzel, A. Weisenburger, G. Mueller, G. Schumacher, A. Jianu, J. Pacio, L. Marocco, R. Stieglitz, T. Wetzel, M. Daubner, K. Litfin, J. B. Vogt, I. Proriol-Serre, D. Gorse, S. Eckert, F. Stefani, D. Buchenau, T. Wondrak, and I. S. Hwang, "Handbook on Lead-bismuth Eutectic Alloy and Lead Properties, Materials Compatibility, Thermal-hydraulics and Technologies - 2015 Edition".

[3] K. Vignarooban, X. Xu, A. Arvay, K. Hsu, and A. M. Kannan, "Heat transfer fluids for concentrating solar power systems – A review", *Applied Energy*, vol. 146, pp. 383–396, 2015.

[4] G. Mueller, W. An, M. DelGiacco, C. Eing, R. Fetzer, B. Flickinger, W. Frey, H. Giese, M. Gottel, C. Gusbeth, A. Heinzel, P. Hoppe, A. Jianu, F. Lang, K. Leber, M. Sack, G. Schuhmacher, J. Singer, R. Straessner, L. Wegner, T. Berghofer, A. Weisenburger, F. Zimmermann, and V. Engelko, "Progress in High Power-particle Beams and Pulsed Power Applications at Karlsruhe Institute of Technology", *Journal of the Korean Physical Society*, vol. 59, no. 61, p. 3588, 2011.

[5] M. Sack, M. Hochberg, and G. Mueller, "Design considerations for a semiconductor-based Marx generator for a pulsed electron beam de-

vice", *Proceedings 2014 International Symposium on Discharges and Electrical Insulation in Vacuum (ISDEIV)*, pp. 381–384, 2014.

[6] V. Engelko, B. Yatsenko, G. Mueller, and H. Bluhm, "Pulsed electron beam facility (GESA) for surface treatment of materials", *Vacuum*, vol. 62, no. 2-3, pp. 211–216, 2001.

[7] V. I. Engelko, V. S. Kuznetsov, and G. Müller, "Electron source of triode type with radial converging electron flow for irradiation of cylindrical targets", *Journal of Applied Physics*, vol. 105, no. 2, p. 23305, 2009.

[8] V. Engelko and G. Mueller, "Influence of particle fluxes from a target on the characteristics of intense electron beams", *Vacuum*, vol. 62, no. 2-3, pp. 97–103, 2001.

[9] G. Mueller, "Application of Pulsed Electron Beams for Improvement of Material Surface Properties", in *BEAMS 2002: 14th International Conference on High-Power Particle Beams*, pp. 325–328, 23-28 June 2002.

[10] V. I. Engelko, "Formation of stable long-pulse electron beams with the help of explosive emission cathodes", *Plasma Devices and Operations*, vol. 13, no. 2, pp. 135–142, 2005.

[11] A. Anders, *Cathodic Arcs: From Fractal Spots to Energetic Condensation*, vol. 50 of *Springer series on atomic, optical, and plasma physics*. New York, NY: Springer-Verlag New York, 2008.

[12] E. A. Litvinov, G. A. Mesyats, and D. I. Proskurovski, "Field emission and explosive electron emission processes in vacuum discharges", *Soviet Physics Uspekhi*, vol. 26, no. 2, pp. 138–159, 1983.

[13] G. A. Mesyats, "Vacuum discharge effects in the diodes of high-current electron accelerators", *IEEE Transactions on Plasma Science*, vol. 19, no. 5, pp. 683–689, 1991.

[14] M. A. Kempkes, R. Phillips, and M. Gaudreau, "Short pulse Marx modulator", in *2015 IEEE International Vacuum Electronics Conference (IVEC)*, IEEE, 2015.

[15] H. Bluhm, *Pulsed Power Systems: Principles and Applications*. Power Systems, Berlin, Heidelberg: Springer-Verlag Berlin Heidelberg, 2006.

[16] F. Martinez-Rodrigo, S. de Pablo, and L. C. Herrero-de Lucas, "Current control of a modular multilevel converter for HVDC applications", *Renewable Energy*, vol. 83, pp. 318–331, 2015.

[17] J. Rodriguez, L. G. Franquelo, S. Kouro, J. I. Leon, R. C. Portillo, M. Prats, and M. A. Perez, "Multilevel Converters: An Enabling Technology for High-Power Applications", *Proceedings of the IEEE*, vol. 97, no. 11, pp. 1786–1817, 2009.

[18] J. Alex and et al., "A New Prototype Modulator for the European XFEL Project in Pulse Step Modulator Technology", *Particle accelerator conference, May 4th - 8th 2009, Vancouver, Canada*, 2009.

[19] E. Marx, "Verfahren zur Schlagpruefung von Isolatoren und anderen elektrischen Vorrichtungen", Patent No. DE455933C, Feb. 13 1923.

[20] G. A. Mesyats, "Matching a line to an explosive-emission diode in local therapy devices", *Technical Physics Letters*, vol. 28, no. 7, pp. 550–551, 2002.

[21] M. Sack and G. Mueller, "Parameter variations for enhanced overvoltage triggering of a switching spark gap", in *2014 ICHVE International Conference on High Voltage Engineering and Application*, IEEE, 2014.

[22] R. L. Cassel, "An all Solid State Pulsed Marx Type Modulator for Magnetrons and Klystrons", in *2005 IEEE Pulsed Power Conference*, pp. 836–838, IEEE, 2005.

[23] M. Hochberg, M. Sack, D. Herzog, A. Weisenburger, W. An, R. Fetzer, and G. Mueller, "A Fast Modular Semiconductor-Based Marx Generator for Driving Dynamic Loads", *IEEE Transactions on Plasma Science*, pp. 627–634, 2018.

[24] L. Redondo, H. Canacsinh, and J. Silva, "Generalized solid-state marx modulator topology", *IEEE Transactions on Dielectrics and Electrical Insulation*, vol. 16, no. 4, pp. 1037–1042, 2009.

[25] W. Jiang, H. Sugiyama, and A. Tokuchi, "Pulsed Power Generation by Solid-State LTD", *IEEE Transactions on Plasma Science*, vol. 42, no. 11, pp. 3603–3608, 2014.

[26] R. J. Adler and V. M. Weeks, "Design and operation of a 700 kV arbitrary waveform generator", in *2009 IEEE Pulsed Power Conference*, pp. 404–407, IEEE, 2009.

[27] R. Phillips, M. Kempkes, M. P. J. Gaudreau, K. Ostlund, and J. Casey, "Affordable, short pulse Marx modulator", *2013 14th International Vacuum Electronics*, 2013.

[28] L. M. Redondo, A. Kandratsyeu, M. J. Barnes, and T. Fowler, "Design strategies for a SiC Marx generator for a kicker magnet", in *2017 IEEE 21st International Conference on Pulsed Power (PPC)*, IEEE, 2017.

[29] A. J. Schwab, *Hochspannungsmesstechnik: Messgeräte und Messverfahren.* Klassiker der Technik, Berlin, Heidelberg: Springer-Verlag Berlin Heidelberg, 2. Aufl., Nachdr. in veränd. Ausstattung der Ausg. Berlin 1981 ed., 2011.

[30] ABB, "Application Note 5SYA 2051", *https://library.e.abb.com/public/6f03cdd0f2264ff48f2992e62497dd5a/Voltage%20ratings%20of%20high%20power%20_%205SYA%202051NLay.pdf, Accessed: Oct. 10 2018.*

[31] C. Felgemacher, S. V. Araujo, C. Noeding, and P. Zacharias, "Benefits of increased cosmic radiation robustness of SiC semiconductors in large power-converters", in *PCIM Europe 2016; International Exhibition and Conference for Power Electronics, Intelligent Motion, Renewable Energy and Energy Management*, pp. 1–8, 2016.

[32] M. Sack, M. Hochberg, and G. Mueller, "Design of a semiconductor-based bipolar Marx generator", *Proceedings 2014 IEEE International Power Modulator and High Voltage Conference (IPMHVC)*, pp. 664–667, 2014,.

[33] M. Sack, M. Hochberg, and G. Mueller, "Synchronized Switching and Active Clamping of IGBT Switches in a Simple Marx Generator", *PCIM*, 2016, Nuremberg, Germany, 10.-12. May 2016.

[34] L. M. Redondo and J. F. Silva, "Repetitive High-Voltage Solid-State Marx Modulator Design for Various Load Conditions", *IEEE Transactions on Plasma Science*, vol. 37, no. 8, pp. 1632–1637, 2009.

[35] H. Canacsinh, J. F. Silva, and L. M. Redondo, "Rise-Time Improvement in Bipolar Pulse Solid-State Marx Modulators", *IEEE Transactions on Plasma Science*, vol. 45, no. 10, pp. 2656–2660, 2017.

[36] I. Benavente and M. Miguez, "A HV pulse generator for PEF applications", 2013.

[37] M. Sack, J. Ruf, M. Hochberg, D. Herzog, and G. Mueller, "A device for combined thermal and pulsed electric field treatment of food", in *2017 International Conference on Optimization of Electrical and Electronic Equipment (OPTIM) & 2017 Intl Aegean Conference on Electrical Machines and Power Electronics (ACEMP)*, pp. 31–36, IEEE, 2017.

[38] A. Schwab and W. Kürner, *Elektromagnetische Verträglichkeit.* Berlin, Heidelberg: Springer Berlin Heidelberg, 2011.

[39] S. Radu, "Engineering Aspects of Electromagnetic Shielding", *SUN Microsystems, http://ieee.rackoneup.net/rrvs/09/Engineering%20Aspects%20of%20 Electromagnetic%20Shielding.pdf, Accessed: Jan. 2018*, 2009.

[40] D. Anke, H.-D. Brüns, B. Deserno, H. Garbe, K.-H. Gonschorek, D. Hansen, P. Harms, J. L. ter Haseborg, S. Keim, A. Kohling, K. Rippl, V. Schmidt, H. Singer, K. H. Gonschorek, and H. Singer, *Elektromagnetische Verträglichkeit.* Wiesbaden: Vieweg+Teubner Verlag, 1992.

[41] G. D. Häberle, *Tabellenbuch Elektrotechnik: Tabellen, Formeln, Normenanwendung*. Europa-Fachbuchreihe für elektrotechnische, elektronische, mechatronische und informationstechnische Berufe, Haan-Gruiten: Verl. Europa-Lehrmittel Nourney Vollmer, 23., neu bearb. und erw. aufl., dr. 2 ed., 2009.

[42] D. Herzog, M. Sack, M. Hochberg, and G. Mueller, "Dynamic Impedance Load for Over-current Testing of Semiconductor-Based

Marx Generators", *2018 IEEE International Power Modulator and High Voltage Conference, 2018, Moran, USA, June 3-7.*

[43] Tyco Electronics Co., "Datasheet IM Relay", *www.te.com, Accessed: Feb. 2017.*

[44] T. Westermann, *Mathematik für Ingenieure*. Berlin, Heidelberg: Springer Berlin Heidelberg, 2015.

[45] M. Beyer, *Hochspannungstechnik: Theoretische und praktische Grundlagen [für die Anwendung]*. Berlin: Springer, berichtigter nachdr ed., 1992.

[46] Philips, "Manual for high-voltage probe P8932", 1978.

[47] HILO TEST, "Datasheet HVT-RCR High voltage dividers", *http://www.hilo-test.de/media/pdf/datasheets/de/HVT-RCR.pdf, Accessd: Feb. 2019.*

[48] M. Sack and G. Mueller, "EMC aspects in the layout of a fast semiconductor switch", *Electromagnetic Compatibility in Automotive Engineering, GMM Specialist Conference*, 2014; Dusseldorf; Germany; 11 March 2014 through 13 March 2014.

[49] LeCroy, "WaveRunner Xi-A Series specifications", *https://teledynelecroy.com/japan/pdf/cata/waverunner_xi-a_spec.pdf, Accessed: Mar. 2019.*

[50] LeCroy, "HDO6000 specifications", *http://cdn.teledynelecroy.com/files/pdf/hdo6000_oscilloscope_datasheet.pdf, Accessed: Mar. 2019.*

[51] M. Hochberg, M. Sack, and G. Mueller, "A Test Environment for Power Semiconductor Devices Using a Gate-Boosting Circuit", *IEEE Transactions on Plasma Science*, vol. 44, no. 10, pp. 2030–2034, 2016.

[52] M. Hochberg, M. Sack, and G. Mueller, "Simple Gate-boosting Circuit for Reduced Switching Losses in Single IGBT Devices", *Proceedings of PCIM 2016, Nuremberg, Germany, 10.-12. May 2016.*

[53] M. Hochberg, M. Sack, and G. Mueller, “Analyzing a Gate-Boosting Circuit for Fast Switching”, *Proceedings 2016 IEEE International Power Modulator and High Voltage Conference (IPMHVC), San Francisco, USA, 05.-09. Jul. 2016.*

[54] Greenleaf Whittier Pickard, “Means For Receiving Intelligence Communicated By Electric Waves”, *US Patent No: US836531, 1906.*

[55] *Halbleiter: Technische Erläuterungen, Technologien und Kenndaten.* Never stop thinking, Erlangen: Publicis Corp. Publ, 3., überarb. und wesentlich erw. aufl. ed., 2004.

[56] J. Millán, “A New Generation of Power Semiconductor Devices”, *SAAEI, July 1st - 3rd 2009, Madrid, Spain*, 2009.

[57] VisIC Technologies, “Datasheet VM40HB120D”, *http://visic-tech.com/wp-content/uploads/2017/08/VM40HB120D-ver1.20.pdf, Accessed: Jan. 2019.*

[58] IXYS RF, “DE475-102N21A Datasheet”, *http://ixapps.ixys.com/datasheet/de475_102n21a.pdf, Accessed: Nov. 2018*, 2009.

[59] Farnell element14, “www.farnell.com”, *Checked Dec. 2018.*

[60] M. Sack, M. Hochberg, and G. Mueller, “Setup and testing of a fast stacked-MOSFET switch”, *2013 IEEE Pulsed Power and Plasma Science Conference (PPPS 2013)*, 2013.

[61] Cree Inc., “Datasheet for SiC MOSFET C2M0080120D”, *http://www.wolfspeed.com/~/media/Files/Cree/Power/Data Sheets/C2M0080120D.pdf, Accessed: Nov. 2015.*

[62] V. K. Khanna, *Insulated gate bipolar transistor (IGBT): Theory and design.* Piscataway, NJ, Hoboken, NJ: Wiley-Interscience, 2003.

[63] J. Lutz, H. Schlangenotto, and U. Scheuermann, *Halbleiter-Leistungsbauelemente: Physik, Eigenschaften, Zuverlässigkeit.* Berlin: Springer, 2006.

[64] A. Volke and M. Hornkamp, *IGBT modules: Technologies, driver and application.* Munich: Infineon Technologies AG, 2011.

[65] I. C. Abdus Sattar, "Insulated Gate Bipolar Transistor (IGBT) Basics, Application Note IXAN0063", *https://www.ixys.com/Documents/AppNotes/IXYS_IGBT_Basic_I.pdf, Accessed Jan. 2019.*

[66] International Recitifier, "IGBT Characteristics, Application Note AN-983", *https://www.infineon.com/dgdl/an-983.pdf?fileId=5546d462533600a40153559f8d921224, Accessed: Jan. 2019.*

[67] ON Semiconductor, "Datasheet for IGBT NGTB40N120IHLWG", *http://www.onsemi.com/pub/Collateral/NGTB40N120IHLW-D.PDF, Accessed: Nov. 2015.*

[68] R. Bayerer and D. Domes, "Parasitic inductance in gate drive circuits", *PCIM Europe Conference Proceedings*, pp. 473–477, 2012.

[69] J. Harper, "Understanding Modern Power MOSFET's", *Fairchild Power Seminar, www.fairchildsemi.com.cn%2FAssets%2FzSystem%2Fdocuments%2Fcollateral%2FonlineSeminars%2FUnderstanding-Modern-Power-MOSFETs-PPT.pdf, Accessed: Jan. 2016*, 2006.

[70] M. Beier-Moebius and J. Lutz, "Breakdown of gate oxide of 1.2 kV SiC-MOSFETs under high temperature and high gate voltage", *Proceedings of PCIM 2016, Nuremberg, Germany, 10.-12. May 2016*, 2016.

[71] I. Cree, "Application Note CPWR-AN25", *https://www.wolfspeed.com/media/downloads/CPWR-AN25.pdf, Accessed: Sep. 2018.*

[72] L. M. Redondo, A. Kandratsyeu, and M. J. Barnes, "Marx Generator Prototype for Kicker Magnets Based on SiC MOSFETs", *IEEE Transactions on Plasma Science*, pp. 3334–3339, 2018.

[73] Y.-C. Jung, "Low loss and high speed IGBT gate driver using the reverse current limiting technique of diode recovery for a hard switching inverter", *International Journal of Electronics*, vol. 81, no. 3, pp. 321–336, 1996.

[74] G. Semiconductor, “Application Note AN-10B: Driving SiC Junction Transistors (SJT): Two-Level Gate Drive Concept”, *https://www.diodemodule.com/wp-content/uploads/2018/06/1305-SJT-AN-10A-Gate-Drive.pdf, Accessed: Jan. 2016.*

[75] M. N. Nguyen, R. L. Cassel, J. E. deLamare, and G. C. Pappas, “Gate drive for high speed, high power IGBTs”, *PPPS-2001: Pulsed Power Plasma Science 2001, Digest of Technical Papers*, 2001.

[76] J. Fu, Z. Zhang, Y.-F. Liu, P. C. Sen, and L. Ge, “A New High Efficiency Current Source Driver With Bipolar Gate Voltage”, *IEEE Transactions on Power Electronics*, vol. 27, no. 2, pp. 985–997, 2012.

[77] Fairchild Semiconductor, “Datasheet FDT86102LZ MOSFET”, *www.fairchildsemi.com/datasheets/FD/FDT86102LZ.pdf, Accessed: Nov. 2015.*

[78] A. Wintrich, U. Nicolai, W. Tursky, and T. Reimann, *Application Manual Power Semiconductors*. Ilmenau: ISLE, 2011.

[79] Young et al., “Process to remove semiconductor chip from a plastic package”, *US Patent No: US09934647, 2002.*

[80] ON Semiconductor, “Datasheet for IGBT HGTG18N120BN”, *https://www.onsemi.com/pub/Collateral/HGTG18N120BN-D.pdf, Accessed: Nov. 2015.*

[81] ON Semiconductor, “Datasheet for IGBT FGA25N120ANTD”, *https://www.onsemi.com/pub/Collateral/FGA25N120ANTDTU-D.PDF, Accessed: Nov. 2015.*

[82] M. Hochberg, M. Sack, D. Herzog, A. Weisenburger, and G. Mueller, “Design Validation of a Single Semiconductor-Based Marx-Generator Stage for Fast Step-Wise Arbitrary Output Waveforms”, *IEEE Transactions on Plasma Science*, vol. 46, no. 10, pp. 3284–3290, 2018.

[83] WIMA, “Datasheet MKP 10 Capacitor series”, *https://www.mouser.de/ datasheet/2/440/e_WIMA_MKP_10-1139811.pdf,*
Accessed: Dec. 2018.

[84] KEMET, "Datasheet C0G Pulse Discharge SMD Capacitors", *https://www.mouser.de/pdfdocs/KEM_C1035_C0G_PULSE_SMD.pdf, Accessed: Dec. 2018*, 2016.

[85] Cornell Dubilier, "Datasheet 944U101K122ACM", *www.cde.com/resources/catalogs/944U.pdf, Accessed: Oct. 2017*, 2015.

[86] ON Semiconductor, "Datasheet RHRG30120", *https://www.onsemi.com/pub/Collateral/RHRG30120-D.PDF, Accessd: Feb. 2019.*

[87] STMicroelectronics, "Datasheet STTH112", *https://www.st.com/resource/en/datasheet/stth112.pdf, Accessed: Feb. 2019.*

[88] Cree Inc., "Datasheet for SiC Diode C4D10120D", *https://www.wolfspeed.com/media/downloads/104/C4D10120D.pdf, Accessed: Jan. 2019.*

[89] ON Semiconductor, "Application Note TND335/D", *https://www.onsemi.com/pub/Collateral/TND335-D.PDF, Accessed: Oct. 2018.*

[90] Fairchild Semiconductor, "Application Note AN-5073: Active Miller Clamp Technology", *https://www.fairchildsemi.com/application-notes/AN/AN-5073.pdf, Accessed: Dec. 2015.*

[91] ON Semiconductor, "Application Note AND9127", *http://www.onsemi.com/pub/Collateral/AND9127-D.PDF, Accessed: Oct. 2018.*

[92] D. Gerber, T. Guillod, R. Leutwyler, and J. Biela, "Gate Unit With Improved Short-Circuit Detection and Turn-Off Capability for 4.5-kV Press-Pack IGBTs Operated at 4-kA Pulse Current", *IEEE Transactions on Plasma Science*, vol. 41, no. 10, pp. 2641–2648, 2013.

[93] Panasonic, "Datasheet for ERJMP4 Metal Plate Resistors", *https://industrial.panasonic.com/cdbs/www-data/pdf/RDN0000/AOA0000C316.pdf, Accessed: Feb. 2019.*

[94] Ohmite, "EY Series Datasheet", *https://www.ohmite.com/assets/docs/res_ey.pdf, Accessed: Jan 2019.*

[95] Broadcom, "Datasheet AFBR-59F2Z", *https://docs.broadcom.com/docs/AV02-4756EN, Accessed: Oct. 2017.*

[96] Broadcom, "Datasheet HFBR-1527ETZ", *https://docs.broadcom.com/docs/AV02-2590EN, Accessed: Feb. 2019.*

[97] O. Veledar, P. O. Byrne, S. Danaher, J. I. H. Allen, L. F. Thompson, and J. E. McMillan, "Simple techniques for generating nanosecond blue light pulses from light emitting diodes", *Measurement Science and Technology*, vol. 18, no. 1, p. 131, 2007.

[98] H. Halbritter, "OSRAM Application Note AN096: High-Speed Switching of IR-LEDs (Part I) Background and Datasheet Definition", *https://www.osram-os.com/Graphics/XPic5/00135349_0.pdf/High-Speed%20Switching%20of%20IR-LEDs.pdf, Accessed: Oct. 24, 2017.*

[99] Analog Devices, "Compensating for the Effects of Input Capacitance on VFB and CFB Op Amps Used in Current-to-Voltage Converters", *https://www.analog.com/media/en/training-seminars/tutorials/MT-059.pdf, Accessed: Jan. 2017.*

[100] Broadcom, "Datasheet HFBR-2526Z", *https://docs.broadcom.com/docs/AV02-2590EN, Accessed: Feb. 2019.*

[101] M. Akiyama, T. Sakugawa, S. H. R. Hosseini, E. Shiraishi, T. Kiyan, and H. Akiyama, "High-Performance Pulsed-Power Generator Controlled by FPGA", *IEEE Transactions on Plasma Science*, vol. 38, no. 10, pp. 2588–2592, 2010.

[102] M. Sack and M. Hochberg, "Internal pulse generator design study", *unpublished*, 2014.

[103] Intel, "Datasheet MAX3000", *https://www.intel.com/content/dam/altera-www/global/en_US/pdfs/literature/ds/m3000a.pdf, Accessed: Feb. 2019.*

[104] M. Sack, J. Sigler, L. Stukenbrock, C. Eing, and G. Mueller, "Lab-scale experiments on PEF-treatment of grape mash at different pulse

parameters", *EAPPC 2012, Sep. 30th - Oct. 4th, Karlsruhe, Germany*, 2012.

[105] Sack M, Attmann F, Müller G, "EMV-Aspekte beim Entwurf einer Elektroporationsanlage", *Presented at: Electromagnetic Compatibility in Automotive Engineering, GMM Specialist Conference;*, pp. 635–642, Dusseldorf; Germany; 09 March 2010 through 11 March 2010.

[106] Wuerth, "WE-SHC SMD Shielding Cabinet Datasheet", *https://katalog.we-online.de/pbs/datasheet/36003500.pdf, Accessed: Jan. 2019.*

Acknowledgments

Throughout this project, I received a lot of support for which I am deeply grateful.

First and foremost, I am deeply grateful to Prof. Dr.-Ing. Georg Müller, who gave me the opportunity to work on this very interesting topic at the Institute for Pulsed Power and Microwave Technology. His great support and trust motivated me to always give my best.
Furthermore, I would like to thank Prof. Dr.-Ing. Thomas Leibfried for kindly accepting to be my second reviewer and the interest he showed in my work.
My deepest gratitude goes to my lab supervisor Priv.-Doz. Dr.-Ing. Martin Sack, who initially promoted me for this project. He shared his comprehensive knowledge of not only pulsed power, but nearly every field of electrical engineering, always asked the right questions and gave me continuous support. Only through his valuable advice and guidance it was possible to complete such a challenging project.
Prof. Dr.-Ing. John Jelonnek and Prof. Dr. Dr. h.c. Manfred Thumm I would like to thank for their valuable advices in the course of this work.
Such a challenging project would of course not have been possible without motivated co-workers. My deepest thanks go to my dearest colleagues B. Eng. Dennis Herzog and Johannes Fleig, who were involved in every stage of this project. They always kept my motivation high and offered help when I most desperately needed it.
With their great knowledge on the GESA device, the GESA team Alfons Weisenburger, Renate Fetzer, Wladimir An, Fabian Lang, Annette Heinzel and Zhen Wang helped me a lot in understanding the device physics. Our many discussions on all sorts of topics were always helpful.
For her valuable contribution in the organization of the 2018 ISP PhD Workshop on Pulsed Power, I would like to express my gratitude to
B. Sc. Melanie Mai.

Additionally, I would like to thank our electronics workshop, namely Klaus Baumann, Kevin Paulus and Robin Grimm for the many fruitful conversations over a cup of coffee and their support on the manufacturing side of the electronic components.
Likewise, I would like to thank our mechanics team, Kai Schäfer, Alexander Sivkovich, Davide Quattrocchi, Hans Brüsemeister and Dennis Kranz for the help in the mechanical design and the setup of the generator.
The great work environment at IHM contributed a lot to my motivation, so I would like to extend my thanks to all colleagues.
During this project, I was lucky to work with many motivated students. Adrian Neukirch, Blasius Forreiter, Ovidiu Popa, Lukasz Watly and Stefan Göbel made a great addition to our team.
Thanks for all the help in administrative matters go to Gabriela Bertsch and Martina Huber.
For her patience, her council and her great support with my work my deep gratitude goes to my partner Tanja Ölschläger – it is done!
My sister Dr. Alexandra Hochberg I would like to thank for her thorough proofreading - I know, I was too slow.
All of this would not have been possible without the continuous and unlimited support of my family and friends. Thank you all!

Awards and publications

Awards

- **Outstanding young researcher award**, EAPPC 2016, Lisbon, Portugal, 18.-22. Sep. 2016
- **Tom R Burkes outstanding graduate student award**, IPMHVC 2016, San Francisco, USA, 05.-09. Jul. 2016

Peer-reviewed publications

- M. Hochberg, M. Sack, D. Herzog, A. Weisenburger, W. An, R. Fetzer and G. Mueller, *A Fast Modular Semiconductor-Based Marx Generator for Driving Dynamic Loads*, in IEEE Transactions on Plasma Science, vol. 47, no. 1, pp. 627-634, 2019.
- G. Loisch, J. Engel, M. Gross, M. Hochberg, H. Huck, G. Koss, O. Lishilin, A. Oppelt, S. Philipp, D. Richter, F. Stephan and P. Weidemann, *Jitter Mitigation in low Density Discharge Plasma Cells for Wakefield Accelerators*, Journal of Applied Physics, vol. 125, no. 6, pp. 063301, 2019
- M. Hochberg, M. Sack, D. Herzog, A. Weisenburger and G. Mueller, *Design Validation of a Single Semiconductor-Based Marx-Generator Stage for Fast Step-Wise Arbitrary Output Waveforms*, in IEEE Transactions on Plasma Science, vol. 46, no. 10, pp. 3284-3290, 2018.
- M. Hochberg, M. Sack and G. Mueller *Analyzing a Gate-Boosting Circuit for Fast Switching*, in Proceedings of the 2016 IEEE International Power Modulator and High Voltage Conference (IPMHVC), pp. 171-175, 2017

- M. Sack, J. Ruf, M. Hochberg, D. Herzog and G. Mueller, *A Device for Combined Thermal and Pulsed Electric Field Treatment of Food*, in Proceedings of the 2017 International Conference on Optimization of Electrical and Electronic Equipment (OPTIM 2017), pp. 31-36, 2017
- M. Hochberg, M. Sack and G. Mueller, *A Test Environment for Power Semiconductor Devices Using a Gate-Boosting Circuit*, in IEEE Transactions on Plasma Science, vol. 44, no. 10, pp. 2030-2034, 2016.
- M. Hochberg, M. Sack and G. Mueller, *Simple Gate-boosting Circuit for Reduced Switching Losses in Single IGBT Devices*, in Proceedings of the PCIM 2016, pp. 1578-1583, Nuremberg, Germany, 10.-12. May 2016
- M. Sack, M. Hochberg and G. Mueller, *Synchronized Switching and Active Clamping of IGBT Switches in a simple Marx Generator*, in Proceedings of PCIM 2016, pp. 1305-1310, 2016
- G. Loisch, M. Gross, H. Huck, A. Oppelt, Y. Renier, F. Stephan, A. Aschikhin, A.M. De La Ossa, J. Osterhoff, M. Hochberg and M. Sack, *A High Transformer Ratio Scheme of PITZ PWFA Experiments*, in Proceedings of the IPAC 2016, pp. 2551-2553, 2016
- M. Sack, M. Hochberg and G. Mueller, *Design of a Semiconductor-based Bipolar Marx Generator*, in Proceedings of the 2014 IEEE International Power Modulator and High Voltage Conference (IPMHVC), pp. 664-667, 2015
- M. Sack, M. Hochberg and G. Mueller, *Design Considerations for a Semiconductor-based Marx Generator for a Pulsed Electron Beam Device*, in Proceedings of the International Symposium on Discharges and Electrical Insulation in Vacuum (ISDEIV), pp. 381-384, 2014
- M. Sack, M. Hochberg and G. Mueller, *Setup and Testing of a fast Stacked-MOSFET switch*, in Digest of Technical Papers - 2013 IEEE Pulsed Power Conference, no. 6627480, 2013
- M. Sack, S. Keipert, M. Hochberg, M. Greule and G. Mueller, *Design Considerations for a fast Stacked-MOSFET switch*, in IEEE Transactions on Plasma Science, vol. 41, no. 10, pp. 2630-2636, 2013

Invited talks

- M. Hochberg, M. Sack, B. Forreiter, W. An, A. Weisenburger, R. Fetzer, D. Herzog, F. Lang and G. Mueller, *Current progress on a fast semiconductor-based Marx generator for a pulsed electron beam device*, Pulsed Power Conference 2017, Brighton, UK, 18.-22. June 2017
- M. Hochberg, M. Sack and G. Mueller, *Pulse Generators and Switching Elements for PEF Treatment Devices*, University of Dunedin, Dunedin, New Zealand, 22.03.2016

Conference contributions

- M. Hochberg, M. Sack, D. Herzog, R. Fetzer, W. An, A. Weisenburger and G. Mueller, *A fast 120 kV, 600 A modular pulsed power source - recent results*, 2nd ISP PhD Workshop, Karlsruhe, Germany, 6.-7. Dec. 2018 (Talk)
- M. Hochberg, M. Sack, A. Weisenburger and G. Mueller, *Modular Semiconductor-based Pulse Power Source for Pulsed Electron Beam Applications*, EAPPC 2018, Changsha, China, 16.-20. Sep. 2018 (Talk)
- M. Hochberg, M. Sack, D. Herzog, A. Weisenburger and G. Mueller, *Current Progress on a fast Semiconductor-based Marx Generator for a Pulsed Electron Beam Device*, ITG Workshop: Vacuum Electronics 2018, Bad Honnef, Germany, 6.-7. Sep. 2018 (Talk)
- M. Hochberg, M. Sack, D. Herzog, F. Lang, W. An, R. Fetzer, A. Weisenburger and G. Mueller, *Development of a 120 kV, 600 A Marx-type power modulator for fast step-wise arbitrary output waveforms*, 2018 IEEE International Power Modulator and High Voltage Conference (IPMHVC), Moran, USA, 3.-7. June 2018 (Talk)
- M. Hochberg, M. Sack, D. Herzog and G. Mueller, *A Fast Modular Pulsed Power Source for a Pulsed Electron Beam Device*, International Symposium on High Voltage Engineering, Buenos Aires, Argentina, 27. Aug. - 01. Sep. 2017 (Talk)

- M. Hochberg, M. Sack, B. Forreiter, W. An, A. Weisenburger, R. Fetzer, D. Herzog, F. Lang and G. Mueller, *Current progress on a fast semiconductor-based Marx generator for a pulsed electron beam device*, Pulsed Power Conference 2017, Brighton, UK, 18.-22. June 2017 (Talk)
- M. Hochberg, M. Sack and G. Mueller, *A Semiconductor-based Marx Generator for a Pulsed Electron beam device*, EAPPC 2016, Lisbon, Portugal, 18.-22. Sep. 2016 (Talk)

KARLSRUHER FORSCHUNGSBERICHTE AUS DEM INSTITUT FÜR HOCHLEISTUNGSIMPULS- UND MIKROWELLENTECHNIK

(ISSN 2192-2764)

Band 1 **MATTHIAS BERINGER**
Design Studies towards a 4 MW 170 GHz Coaxial-Cavity Gyrotron. 2011
ISBN 978-3-86644-663-2

Band 2 **JENS FLAMM**
Diffraction and Scattering in Launchers of
Quasi-Optical Mode Converters for Gyrotrons. 2012
ISBN 978-3-86644-822-3

Band 3 **MATTIA DEL GIACCO**
Investigation of Fretting Wear of Cladding Materials in Liquid Lead. 2013
ISBN 978-3-86644-960-2

Band 4 **AMITAVO ROY CHOUDHURY**
Investigations of After Cavity Interaction in Gyrotrons
Including the Effect of Non-uniform Magnetic Field. 2013
ISBN 978-3-7315-0129-9

Band 5 **MICHAEL BETZ**
The CERN Resonant WISP Search (CROWS). 2014
ISBN 978-3-7315-0199-2

Band 6 **ANDREAS SCHLAICH**
Time-dependent spectrum analysis of high power gyrotrons. 2015
ISBN 978-3-7315-0375-0

Band 7 **DHIDIK PRASTIYANTO**
Temperature- and Time-Dependent Dielectric Measurements
and Modelling on Curing of Polymer Composites. 2016
ISBN 978-3-7315-0424-5

Band 8 **YIMING SUN**
Adaptive and Intelligent Temperature Control of Microwave
Heating Systems with Multiple Sources. 2016
ISBN 978-3-7315-0467-2

Die Bände sind unter www.ksp.kit.edu als PDF frei verfügbar oder als Druckausgabe bestellbar.

Band 9 **JIANGHUA ZHANG**
Influence of Emitter surface roughness and Emission inhomogeneity on Efficiency and stability of high power Fusion gyrotrons. 2016
ISBN 978-3-7315-0578-5

Band 10 **ANTON MALYGIN**
Design and Experimental Investigation of a Second Harmonic 20 kW Class 28 GHz Gyrotron for Evaluation of New Emitter Technologies. 2016
ISBN 978-3-7315-0584-6

Band 11 **JOACHIM FRANCK**
Systematic Study of Key Components for a Coaxial-Cavity Gyrotron for DEMO. 2017
ISBN 978-3-7315-0652-2

Band 12 **PARTH CHANDULAL KALARIA**
Feasibility and Operational Limits for a 236 GHz Hollow-Cavity Gyrotron for DEMO. 2017
ISBN 978-3-7315-0717-8

Band 13 **CHUANREN WU**
Conceptual Studies of Multistage Depressed Collectors for Gyrotrons. 2019
ISBN 978-3-7315-0934-9

Band 14 **MARTIN HOCHBERG**
A high-voltage pulsed power modulator for fast-rising arbitrary waveforms. 2020
ISBN 978-3-7315-0958-5